__Disclaimer__

The publisher of this book is by no way associated with the National Institute of Standards and Technology (NIST). The NIST did not publish this book. It was published by 50 page publications under the public domain license.

50 Page Publications.

Book Title: Face Recognition Vendor Test 2006 and Iris Challenge Evaluation 2006 Large-Scale Results

Book Author: P J. Phillips; K W. Bowyer; P J. Flynn; Alice J. O'Toole; W T. Scruggs; Cathy L. Schott; Matthew Sharpe;

Book Abstract: The Face Recognition Vendor Test (FRVT) 2006 and Iris Challenge Evaluation (ICE) 2006 are independent U.S. Government evaluations of face and iris recognition performance. These evaluations were conducted simultaneously at NIST using the same test infrastructure. Human performance verses algorithm performance was also evaluated, making this the first unified evaluation of face, iris and human recognition performance in the U.S. Government. Test results on large-scale experiments for both face and iris recognition are presented in this evaluation report. These results show two orders of magnitude improvement in performance on face recognition since the first face recognition evaluation program was established in 1993 and provide a performance baseline for iris recognition. Results on human performance verses machine performance and the first multi-modal comparison between face and iris recognition are also presented in this report.

Citation: NIST Interagency/Internal Report (NISTIR) - 7408

Keyword: biometrics;biometrics evaluation;face recognition;Face Recognition Vendor Test;FRVT, FRVT 2006;Iris Challenge Evaluation, ICE 2006;iris recognition

FRVT 2006 and ICE 2006 Large-Scale Results

March 2007

P. Jonathon Phillips[1], W. Todd Scruggs[2], Alice J. O'Toole[3], Patrick J. Flynn[4], Kevin W. Bowyer[4], Cathy L. Schott[5], Matthew Sharpe[2]

[1]National Institute of Standards and Technology, 100 Bureau Dr., Gaithersburg, MD 20899

[2]SAIC, 4001 N. Fairfax Dr., Arlington, VA 22203

[3]School of Behavioral and Brain Sciences, GR4.1, The U. of Texas at Dallas, Richardson, TX 75083-0688

[4]Computer Science & Engineering Depart., U. of Notre Dame, Notre Dame, IN 46556

[5]Schafer Corp., 4601 N. Fairfax Drive, Suite 1150, Arlington, VA 22203

NISTIR 7408

National Institute of Standards and Technology
Gaithersburg, MD 20899

Sponsors

- Department of Homeland Security
  - Science and Technology Department
  - Transportation Security Administration

- Director of National Intelligence
  - Information Technology Innovation Center

- Federal Bureau of Investigation

- National Institute of Justice

- Technical Support Working Group

FRVT 2006 and ICE 2006 Large-Scale Results[*]

P. Jonathon Phillips[1], W. Todd Scruggs[2], Alice J. O'Toole[3], Patrick J. Flynn[4]
Kevin W. Bowyer[4], Cathy L. Schott[5], Matthew Sharpe[2]

[1]National Institute of Standards and Technology, 100 Bureau Dr., Gaithersburg, MD 20899
[2]SAIC, 4001 N. Fairfax Dr., Arlington, VA 22203
[3]School of Behavioral and Brain Sciences, GR4.1, The U. of Texas at Dallas, Richardson, TX 75083-0688
[4]Computer Science & Engineering Depart., U. of Notre Dame, Notre Dame, IN 46556
[5]Schafer Corp., 4601 N. Fairfax Drive, Suite 1150, Arlington, VA 22203

March 29, 2007

Abstract

This report describes the large-scale experimental results from the Face Recognition Vendor Test (FRVT) 2006 and the Iris Challenge Evaluation (ICE) 2006. The FRVT 2006 looks at recognition from high-resolution still images and three-dimensional (3D) face images, and measures performance for still images taken under controlled and uncontrolled illumination. The ICE 2006 reports iris recognition performance from left and right iris images. The FRVT 2006 results from controlled still images and 3D images document an order-of-magnitude improvement in recognition performance over the FRVT 2002. This order-of-magnitude improvement was one of the goals of the preceding technology development effort, the Face Recognition Grand Challenge (FRGC). The FRVT 2006 and the ICE 2006 compared recognition performance from very-high resolution still face images, 3D face images, and single-iris images. On the FRVT 2006 and the ICE 2006 datasets, recognition performance was comparable for all three biometrics. In an experiment comparing human and algorithm performance, the best-performing face recognition algorithms were more accurate than humans. These and other results are discussed in detail.

---

[*]Please direct correspondence to P. Jonathon Phillips at jonathon@nist.gov. We acknowledge the support of Department of Homeland Security's Science and Technology Department and Transportation Security Administration (TSA), the Director of National Intelligence's Information Technology Innovation Center, the Federal Bureau of Investigation (FBI), the National Institute of Justice, and the Technical Support Working Group (TSWG). The identification of any commercial product or trade name does not imply endorsement or recommendation by the National Institute of Standards and Technology, SAIC, Schafer Corp., U. of Texas at Dallas or U. of Notre Dame.

1 Introduction

The Face Recognition Vendor Test (FRVT) 2006 and the Iris Challenge Evaluation (ICE) 2006 are evaluations of face and iris recognition technology, respectively. A technology evaluation evaluates the performance of the underlying technology [1]. A technology evaluation is different from a scenario evaluation, which assesses how well a biometric technology meets the requirements for a particular scenario. The design of both the FRVT 2006 and ICE 2006 shares the same protocol and they report results on biometric samples from a multi-biometric dataset. Both evaluations together constitute the first multi-biometric technology evaluation that measures performance on iris recognition technology, and still and three-dimensional (3D) face recognition techniques.

Face and iris are two biometrics that have been developed over the last 20 years. Face recognition is a vibrant area of biometrics with active research and commercial efforts [2]. The FRVT 2006 is the the latest in a series of evaluations for face recognition that began in 1993. With the expiration of the Flom and Safir [3] iris recognition patent in 2005, iris recognition algorithm development has become more active [4]. The ICE 2006 is the first independent evaluation for iris recognition algorithms. Since face and iris are competitive and complementary biometric technologies, conducting a simultaneous technology evaluation allowed for assessments of each biometric and comparison of their capabilities.

The key results and accomplishments of the FRVT 2006 and the ICE 2006 are:

- The FRVT 2006 and the ICE 2006 established the first independent performance benchmark for iris recognition technology and 3D face recognition technology. These two benchmarks enable (for the first time) the identification of the most promising technological approaches for the respective biometrics. These benchmarks also allow progress to be measured.

- FRVT 2006 and ICE 2006 are the first technology evaluations that allowed iris recognition, still face recognition, and 3D face recognition performance to be compared. The results on the multi-biometric dataset show that the performance for all three biometrics is comparable. The combination of the FRVT 2006 and the ICE 2006 constitutes the first technology evaluation designed to measure and compare the performance of multiple biometrics.

- The FRGC was a face recognition technology development effort that supported the development of the face recognition algorithms from high-resolution still and 3D imagery [5][6]. The goal of the Face Recognition Grand Challenge (FRGC) was a decrease in the error rate of face recognition algorithms by an order of magnitude. The FRVT 2006 documented a decrease in the error rate by at least an order of magnitude over what was observed in the FRVT 2002 [7]. This decrease in error rate was achieved by still and by 3D face recognition algorithms.

- The FRVT 2006 documented significant progress since January 2005 in face recognition when faces are matched across different lighting conditions. In the FRVT 2006, which was an evaluation on sequestered data, five submissions performed better than the best results in the January 2005 FRGC results [6]. The observed increase occurred despite the FRGC being an open challenge problem with the identities of faces known to the FRGC participants and the FRVT 2006 being a sequestered evaluation.

- For the first time in a biometric evaluation, the FRVT 2006 directly compared human and machine face recognition performance. The results show that, at low false alarm rates for humans, seven automatic face recognition algorithms were comparable to or better than humans at recognizing faces taken under different lighting conditions. Furthermore, three of the seven algorithms were comparable to or better than humans for the full range of false alarm rates measured.

The FRVT 2006 and the ICE 2006 results in this report support the claims above. The report is organized as follows. Sections 2 and 3 provide background and overview material for the two evaluations. Prior to discussing the multi-biometric aspects of the evaluation, the experimental results for both the individual biometrics (iris and face) are presented. Section 4 presents the ICE 2006 results, and Section 5 presents the FRVT 2006 results. In Section 5, the still portion of the FRVT 2006, including human performance, is discussed first, followed by the 3D face recognition benchmark. The multi-biometric aspects of the ICE

2006 and the FRVT 2006 are discussed in section 6 and overall conclusions are presented and discussed in section 7. Appendix A-1 describes the method and materials, and Appendix A-2 gives detailed performance results.

2 Progress in Iris and Face Recognition

The idea of the iris as a biometric is relatively new, with Flom and Safir patenting the concept in 1987 [3]. The seminal work of Daugman on automatic iris recognition was published in 1993 [8]. The Daugman algorithm was the basis for the first commercial iris recognition system. Because of the Flom and Safir patent and the lack of a publicly accessible dataset of images, there was limited research in iris recognition for most of the decade following the publication of Daugman's algorithm. With the expiration of the Flom and Safir patent, and the availability of the CASIA dataset and the ICE 2005 challenge problem and dataset, research activity in iris recognition has greatly increased in recent years. A recent evaluation of iris performance was ITIRT, which measured interoperability of the Iridian algorithm for four sensors [9].

The face recognition community has benefited from a series of U.S. Government funded technology development efforts and evaluation cycles, beginning with the FERET program in September 1993. One of the key contributions and legacies of these development efforts is the large data sets collected to support these efforts. The large datasets have spurred the development of new algorithms. The independent evaluations have provided an unbiased assessment of the state-of-the-art in the technology and have identified the most promising approaches. In addition, the evaluations have documented two orders of magnitude improvement in performance from the start of the FERET program through the FRVT 2006.

Figure 1 quantifies this improvement at four key milestones. For each milestone, the false reject rate (FRR) at a false accept rate (FAR) of 0.001 (1 in 1000) is given for a representative state-of-the-art algorithm. The 1993 milestone is a retrospective implementation of Turk and Pentland's eigenface algorithm [10], which was partially automatic (it required that eye coordinates be provided). Performance is reported on the eigenface implementation of Moon and Phillips [11] with the FERET Sept96 protocol [12], in which images of

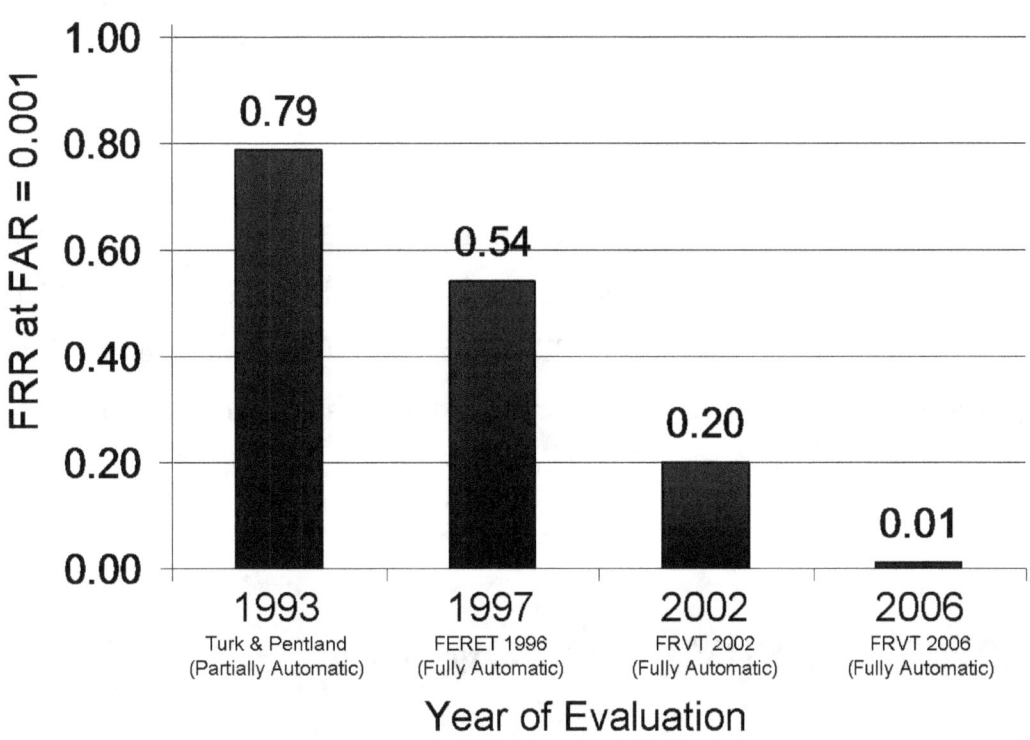

Figure 1: THE REDUCTION IN ERROR RATE FOR STATE-OF-THE-ART FACE RECOGNITION ALGORITHMS AS DOCUMENTED THROUGH THE FERET, THE FRVT 2002, AND THE FRVT 2006 EVALUATIONS.

a subject were taken on different days (dup I probe set). The 1997 milestone is for the Sept97 FERET evaluation, which was conducted at the conclusion of the FERET program. Performance is quoted on the U. of Southern California's fully automatic submission to the final FERET evaluation [13][14]. The 1993 and 1997 results are on the same test dataset and show improvement in algorithm technology under the FERET program. Technology improved from partially automatic to fully automatic algorithms, while error rate declined by approximately a third.

The 2002 benchmark is from the FRVT 2002. Verification performance is reported for the Cognitec, Eyematic, and Identix submissions on the low resolution facial image dataset. The 2006 benchmark is from the FRVT 2006. Here, a FRR of 0.01 at a FAR of 0.001 was achieved by Neven Vision (NV1-NORM algorithm) on the very high-resolution still images and Viisage (V-3D-N algorithm) on the 3D images. Both sets of images were from the multi-biometrics dataset. The improvement in algorithm performance between FRVT 2002 and FRVT 2006 is due to advancement in algorithm design, sensors, and understanding of the importance of correcting for varying illumination across images.

One key factor in the rapid reduction in the error rate over 13 years was the U.S Government sponsored evaluations and challenge problems. The FERET and the FRGC challenge problems focused the research community on large datasets and challenge problems designed to advanced face recognition technology. The FERET, the FRVT 2002 and the FRVT 2006 evaluations provided performance benchmarks, measured progress of, and assessed the state of the underlying technology with the goal of providing researchers with feedback on the efficacy of their approaches.

3 ICE 2006 and FRVT 2006 Overview

The FRVT 2006 and the ICE 2006 were independent evaluations. Both evaluations were open to universities, research institutes, and companies. Participants submitted algorithms to NIST for evaluation. These algorithms were tested on data which was sequestered at the subject level; i.e., biometric samples of the subjects in the ICE 2006 and the FRVT 2006 had not been previously released.

The FRVT 2006 measured performance on single frontal facial imagery and three dimensional (3D) face scans. In the FRVT 2006, a 3D face scan consisted of a texture and a shape channel, see Figure 2. The ICE 2006 measured performance on both right and left irises.

Results are reported on three datasets. The first is the *multi-biometric dataset*, which consists of very-high resolution still frontal facial images (referred to as the *very-high resolution* dataset), three-dimensional (3D) facial scans (referred to as the *3D* dataset), and iris images, see Figure 2. The very-high resolution images were taken with a 6 Mega-pixel Nikon D70 camera, the 3D images with a Minolta Vivid 900/910 sensor, and the iris images with a LG EOU 2200. The images obtained using the LG EOU 2200 for the ICE 2006 evaluations intentionally represent a broader range of quality than the sensor would normally acquire, see Appendix A-1. In the multi-biometric dataset, biometric samples for all three biometrics were collected from the same subject pool. The very-high resolution still images in the multi-biometric data set were collected under controlled and uncontrolled illumination conditions. The average face size for the controlled images was 400 pixels between the centers of the eyes and 190 pixels for the uncontrolled images. The 3D and iris data were collected with active illumination source. Both active illuminations were an integral part of the sensor. The second dataset is the *high-resolution dataset*, which consisted of high resolution frontal facial images taken under both controlled and uncontrolled illumination. The high-resolution images were taken with a 4 Megapixel Canon PowerShot G2. The average face size for the controlled images was 350 pixels between the centers of the eyes and 110 pixels for the uncontrolled images. The third is the *low resolution dataset*, consisting of low resolution images taken under controlled illumination conditions. The low-resolution dataset is the same dataset used in the HCInt portion of the FRVT 2002. The low-resolution images were JPEG compressed to a storage size on disk of approximately 10,000 bytes with an average face size of 75 pixels between the centers of the eyes.

The ICE 2006 measured performance for one-to-one matching algorithms and the FRVT 2006 measured performance for both one-to-one and normalization matching algorithms[1]. In one-to-one matching, the comparison of two biometric samples is solely a function of the two samples; i.e., a one-to-one algorithm will return the same match score between two biometric samples regardless of whether the samples are compared independently or as part of a large set. In normalized matching, and algorithm will adjust the internal face representation based on the face images in the gallery (enrolled dataset). Normalization generally improves performance.

Assessing the state-of-the-art of face and iris recognition technology required measuring performance for fully automatic algorithms on large-scale experiments. For the ICE 2006 submissions, analysis was restricted to algorithms that could complete the large-scale iris experiments in three weeks of processing time on a single Intel Pentium 4 3.6GHz 660 processor. For the FRVT 2006 submissions, the large-scale requirements translated into restricting the analysis to fully automated systems; i.e., algorithms that do not need eye

---

[1] Normalization can also be referred to as cohort or gallery normalization.

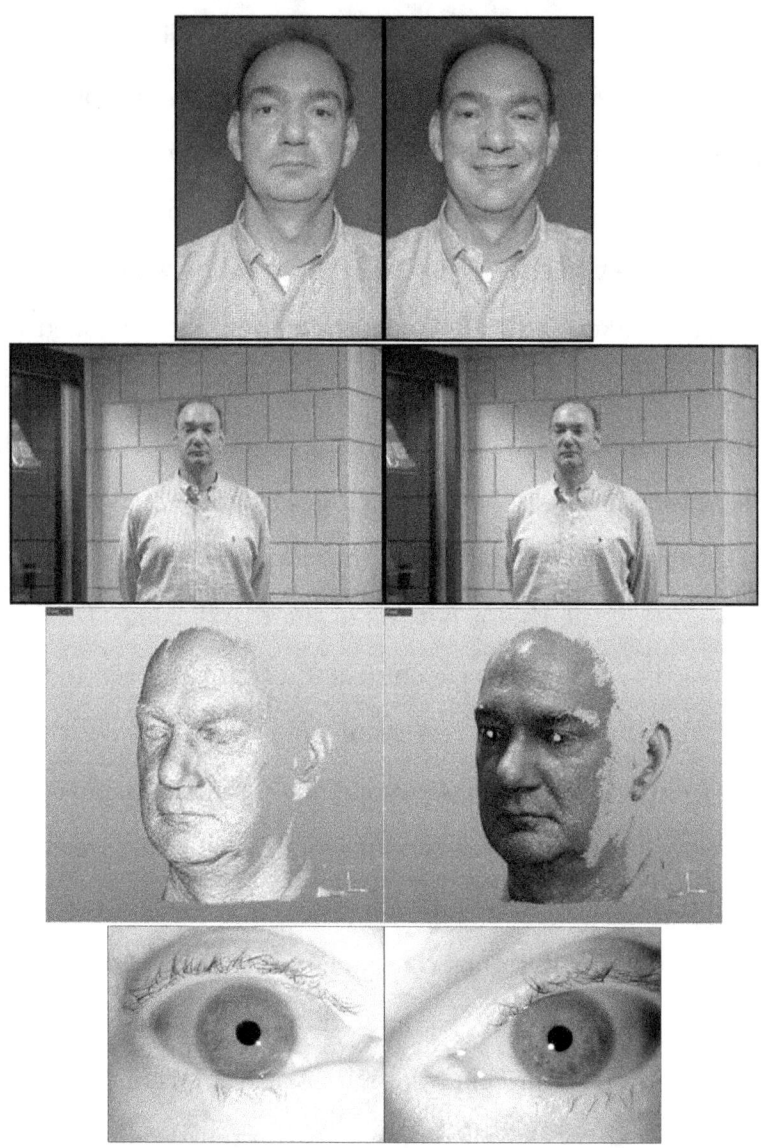

Figure 2: An example of the types of images used in FRVT 2006 and ICE 2006. The top row shows two frontal images taken under controlled illumination with neutral and smiling expressions. The second row shows two images taken under uncontrolled illumination with neutral and smiling expressions. The third contains a 3D facial image. The left images is the shape channel only and the right image has the texture channel on top of the shape channel. The bottom row contains right and left iris images. All samples are from the multi-biometrics dataset.

coordinates or other auxiliary meta-data. Because the ICE 2006 and the FRVT 2006 are technology evaluations, the performance results presented in the main body of this report directly support the assessment of the state-of-the-art. Table 1 lists the algorithms used in the large-scale analysis. Appendix A-2 covers detailed performance for all algorithms that completed the large-scale experiments.

Table 1: THE LIST OF ALGORITHMS COVERED IN THE LARGE SCALE ANALYSIS. COLUMN HEADINGS IDENTIFY EACH PARTICIPANT GROUP AND FIVE BIOMETRIC MATCHING TASK IN THE FRVT 2006 AND THE ICE 2006. THE ORGANIZATION THAT SUBMITTED AN ALGORITHM IS LISTED IN THE GROUP COLUMN. THE ABBREVIATIONS USED IN THE FIGURES ARE PRESENTED IN THE TABLE. A BLANK CELL IN A COLUMN FOR A GROUP MEANS THEY DID NOT SUBMIT AN ALGORITHM FOR THE TASK IN THAT COLUMN.

Group	Iris	Still 1to1	Still norm	3D 1to1	3D norm	Shape
U. of Cambridge	CAM-2					
Cognitec		COG1-1TO1	COG1-NORM	COG1-3D	COG1-3D-N	
Geometrix						GEO-SH
U. of Houston						HO3-SH
Identix		IDX4-1TO1	IDX1-NORM			
Iritech	IRTCH-2					
Neven Vision		NV1-1TO1	NV1-NORM			
Rafael		RA-1TO1	RA-NORM			
Sagem		SG2-1TO1	SG2-NORM			
Sagem-Iridian	SI-2					
SAIT		ST-1TO1	ST-NORM			
Toshiba		TO2-1TO1	TO1-NORM			
Tsinghua U.		TS2-1TO1	TS2-NORM	TS1-3D		
Viisage		V-1TO1	V-NORM	V-3D	V-3D-N	

4 ICE 2006

The ICE 2006 establishes the first independent performance benchmark for iris recognition algorithms. Performance for the ICE 2006 benchmark is presented in Figure 3 for algorithms from three groups: Sagem-Iridian (SG-2), Iritech (IRTCH-2), and Cambridge (CAM-2), see Figure 4 for an explanation of boxplots. The interquartile range for all three algorithms overlaps with the largest amount of overlap between Iritech (IRTCH-2), and Cambridge

(CAM-2). Over all three algorithms, the smallest interquartile is a FRR of 0.09 at a FAR of 0.001 and the largest interquartile is a FRR of 0.26 at a FAR of 0.001.

The results in the ICE 2005, a technology development effort, showed that for the top four groups, recognition performance on the right eye was better than the left eye. In the ICE 2006, the median FRR for the left eye is always smaller than the median FRR for the right eye; however, the range of the boxplots is similar. The results of the ICE 2006 show the same relative performance level. This is seen in Figure 3 by the range of the boxplots for all three algorithms. Hence, the difference in performance observed in ICE 2005 was not confirmed by the results in the ICE 2006. The difference between the ICE 2005 and the ICE 2006 conclusions may be because of the smaller number of samples in the ICE 2005 than the ICE 2006 (2953 versus 59,558) and because the ICE 2005 characterized performance for each eye by one partition versus 30 partitions for each eye in the ICE 2006.

The execution time varied significantly between the Cambridge submission and the Sagem-Iridian and Iritech submissions. The Cambridge algorithm (CAM-2) took 6 hours to complete the ICE 2006 large scale experiments and the Sagem-Iridian (SI-2) algorithm and Iritech (IRTCH-2) algorithms took approximately 300 hours.

The performance of a biometric system will vary with different sets of biometric samples. This is true even when biometric samples are taken under the same conditions; e.g., in face recognition, matching images taken under controlled illumination. It is important to measure both the overall performance of a biometric system and the scale of the variability to measure statistical uncertainty. In the ICE 2006 and the FRVT 2006, performance variability is measured by partitioning the test images into a set of smaller test sets. Performance is then computed on each of the partitions. For example, the test image set for the right iris consists of 29,056 images of 240 subjects. This test set was divided into 30 smaller test sets. Performance for each partition was computed for each algorithm[2]. Performance over a set of partitions is reported graphically on a boxplot, see Figure 4.

---

[2] Details on the partitioning and scoring protocol is given in Appendix A-1.

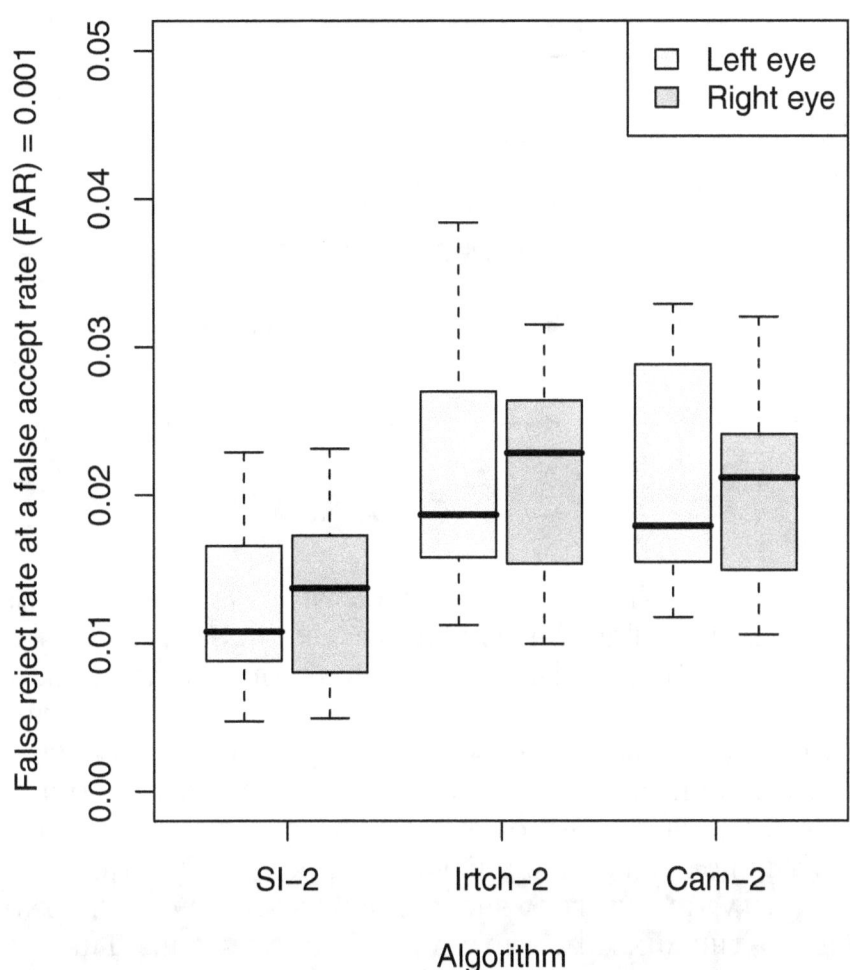

Figure 3: SUMMARY PERFORMANCE OF THE ICE 2006. RESULTS ARE PRESENTED FOR THREE GROUPS: CAMBRIDGE (CAM-2), IRITECH (IRTCH-2) AND SAGEM-IRIDIAN (SI-2). PERFORMANCE IS BROKEN OUT BY RIGHT AND LEFT EYES. THE FALSE REJECT RATE (FRR) AT A FALSE ACCEPT RATE (FAR) OF 0.001 IS REPORTED. PERFORMANCE IS REPORTED FOR 29,056 RIGHT AND 30,502 LEFT IRIS IMAGES FROM 240 SUBJECTS WITH 30 PARTITIONS FOR EACH EYE.

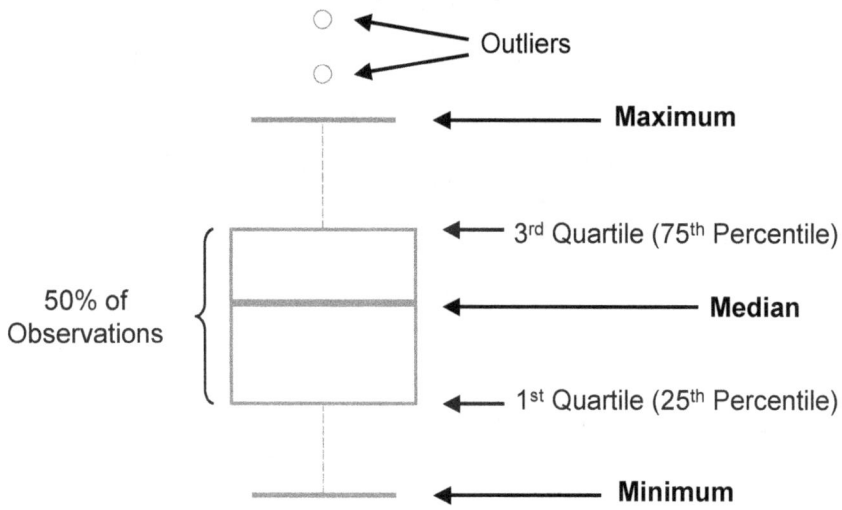

Figure 4: AN EXAMPLE OF A BOXPLOT WITH LOCATION OF DESCRIPTIVE STATISTICS LABELED. THE HORIZONTAL LINE THROUGH THE MIDDLE OF THE BOX IS THE MEDIAN OF THE PERFORMANCE RANGE (50% OF THE OBSERVATIONS ARE GREATER THAN THE MEDIAN AND 50% ARE LESS THAN THE MEDIAN). THE TOP AND BOTTOM OF THE BOX MARKS THE 1st QUARTILE (25th PERCENTILE) AND 3rd QUARTILE (75th PERCENTILE) VALUES OF THE OBSERVATIONS RESPECTIVELY. (AT THE 25th PERCENTILE POINT, 25% OF THE DATA HAS VALUES LESS THAN THIS POINT.) THUS, 50% OF THE PERFORMANCE RANGE IS CONTAINED IN THE BOX. ABOVE AND BELOW THE BOX ARE VERTICAL DASHED LINES, THE "WHISKERS", THAT END WITH A SHORT HORIZONTAL LINE. THE ENDS OF WHISKERS CORRESPOND TO THE MINIMUM AND MAXIMUM DATA VALUE. THE CIRCLES ABOVE OR BELOW THE WHISKERS REPRESENT OUTLIERS. (TO BE TECHNICALLY ACCURATE, THE LENGTH OF THE WHISKER IS THE SMALLER OF THE MAXIMUM MINUS THE 3rd QUARTILE THE (OR THE 1st QUARTILE MINUS THE MINIMUM) AND 1.5 TIMES THE VERTICAL DIMENSION OF THE BOX.)

5 FRVT 2006

The FRVT 2006 large-scale experiments documented progress in face recognition in four areas. First, the FRGC goal of improving performance by an order of magnitude over FRVT 2002 was achieved. Second, the FRVT 2006 established the first 3D face recognition benchmark. Third, the FRVT 2006 showed significant progress has been made in matching faces across changes in lighting. Fourth, the FRVT 2006 showed that face recognition algorithms are capable of performing better than humans.

5.1 Controlled Illumination

The goal of the FRGC was to improve face recognition performance to achieve a FRR of 0.02 at a FAR of 0.001 for matching facial images taken under controlled illumination. This goal was exceeded on the FRVT 2006 very-high resolution dataset with algorithms achieving a FRR of 0.01.

Figure 5 summarizes performance of face recognition for still images under controlled illumination for three datasets: very-high resolution, high resolution, and low resolution. On the very-high resolution dataset, four algorithms met or exceeded the FRGC goal of a FRR of 0.02. These algorithms are from Neven Vision (NV1-NORM and NV1-1TO1[3]), Viisage (V-NORM) and Cognitec (COG1-NORM). On the high resolution dataset, the Neven Vision (NV1-NORM) algorithm with a FRR interquartile range of 0.021 to 0.023 came close to meeting the FRGC goal.

Three algorithms on the very-high resolution dataset had performance that crossed the FRR of 0.01 at a FAR of 0.001 threshold. The FRR interquartile range for the three algorithms are 0.006 to 0.015 for NV1-NORM, 0.008 to 0.016 for NV1-1TO1, and 0.010 to 0.017 for V-NORM.

The best performer on the low-resolution dataset at FAR=0.001 was Toshiba (TO1-NORM) with an interquartile FRR range of 0.024 to 0.027. Four algorithms, Neven Vision (NV1-NORM), Viisage (V-NORM), Cognitec (COG1-NORM), and Sagem (SG2-NORM) had performance in the neighborhood of FRR = 0.05 at a FAR of 0.001. The lowest quartile

---

[3]The algorithm NV1-1TO1 was not plotted on Figure 5.

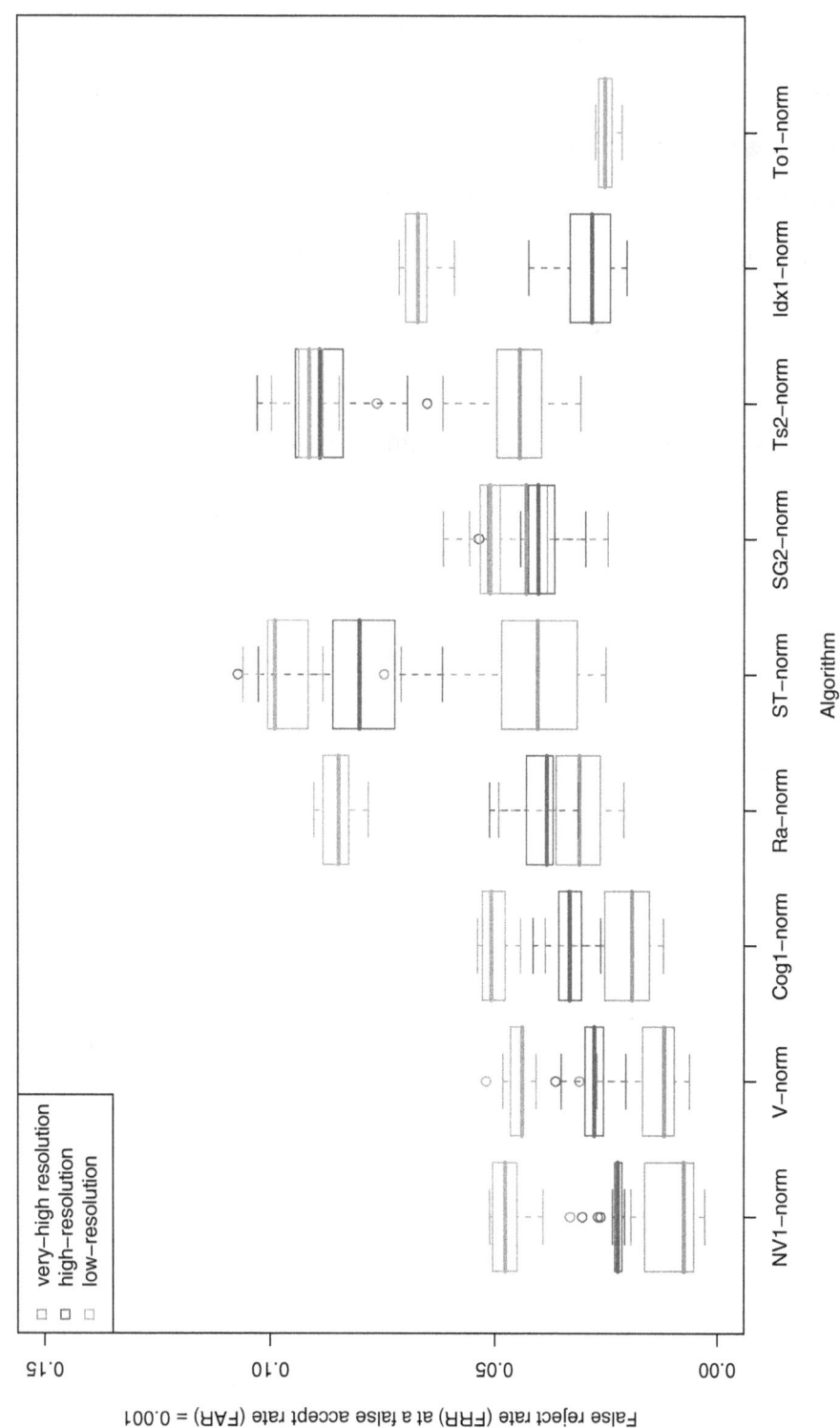

Figure 5: SUMMARY OF STILL FACE RECOGNITION PERFORMANCE ON THE VERY HIGH-RESOLUTION, HIGH-RESOLUTION, AND LOW-RESOLUTION DATA SETS. EACH COLUMN IN THE GRAPH REPORTS PERFORMANCE FOR ONE ALGORITHM WITH RESULTS PROVIDED FOR UP TO THREE DATA SETS. FOR EACH ALGORITHM, THE PERFORMANCE RESULTS ON A DATA SET ARE REPORTED BY A DIFFERENT COLOR BOXPLOT. FOR A SAGEM (SG2-NORM) ALGORITHM, THE BODY OF THE BOXPLOTS OVERLAP FOR ALL THREE DATASETS. FOR A TSINGHUA (TS2-NORM) ALGORITHM, THE BODY OF THE BOXPLOTS OVERLAP THE HIGH-RESOLUTION AND LOW-RESOLUTION DATASETS. FOR IDENTIX (IDX1-NORM) AND TOSHIBA (TO1-NORM), PERFORMANCE WAS OUTSIDE THE RANGE OF THIS GRAPH FOR AT LEAST ONE DATASET.

from this grouping was a FRR of 0.043 and the highest was a FRR of 0.053. While Toshiba performed extremely well on the low-resolution data set at FAR=0.01 and FAR=0.001, their performance was not consistent across all the still datasets.

For the four algorithms Neven Vision (NV1-NORM), Viisage (V-NORM), Cognitec (COG1-NORM), and SAIT (ST-NORM), there is a clear ranking of the difficulty of the three datasets, with the low-resolution being the most difficult and the very-high resolution dataset being easiest; i.e., having the best performance. The primary difference between the three datasets is the size of the faces and consistency of the lighting.

5.2 3D Face Recognition

The FRVT 2006 provides the first benchmarks of 3D face recognition technology. Benchmarks are provided for one-to-one and normalization approaches that use both shape and texture, and for one-to-one shape only techniques. Performance for 3D face recognition is summarized in Figure 6. All results are from the 3D portion of the multi-biometric dataset.

Performance on the 3D dataset meets the FRGC goal of an order of magnitude improvement in performance. The best performers for 3D have a FRR interquartile range of 0.005 to 0.015 at a FAR of 0.001 for the Viisage normalization (V-3D-N) algorithm and a FRR interquartile range of 0.016 to 0.031 at a FAR of 0.001 for the Viisage 3D one-to-one (V-3D) algorithm. Both algorithms met the FRGC performance goal. The shape only benchmark was set by the Geometrix (GEO-SH) and the U. of Houston (HO3-SH) submissions.

On the FRVT 3D dataset, the normalized algorithms performed better than the one-to-one algorithms. This is seen by comparing the results for the Cognitec and Viisage 3D normalized algorithms (COG1-3D-N and V-3D-N) to their counterpart one-to-one algorithms (V-3D and V-3D).

5.3 Uncontrolled Illumination

When compared with the FRGC results, the FRVT 2006 shows a significant improvement in recognition when matching faces across changes in lighting. In these experiments, the enrolled images are frontal facial images taken under *controlled* illumination and the probe

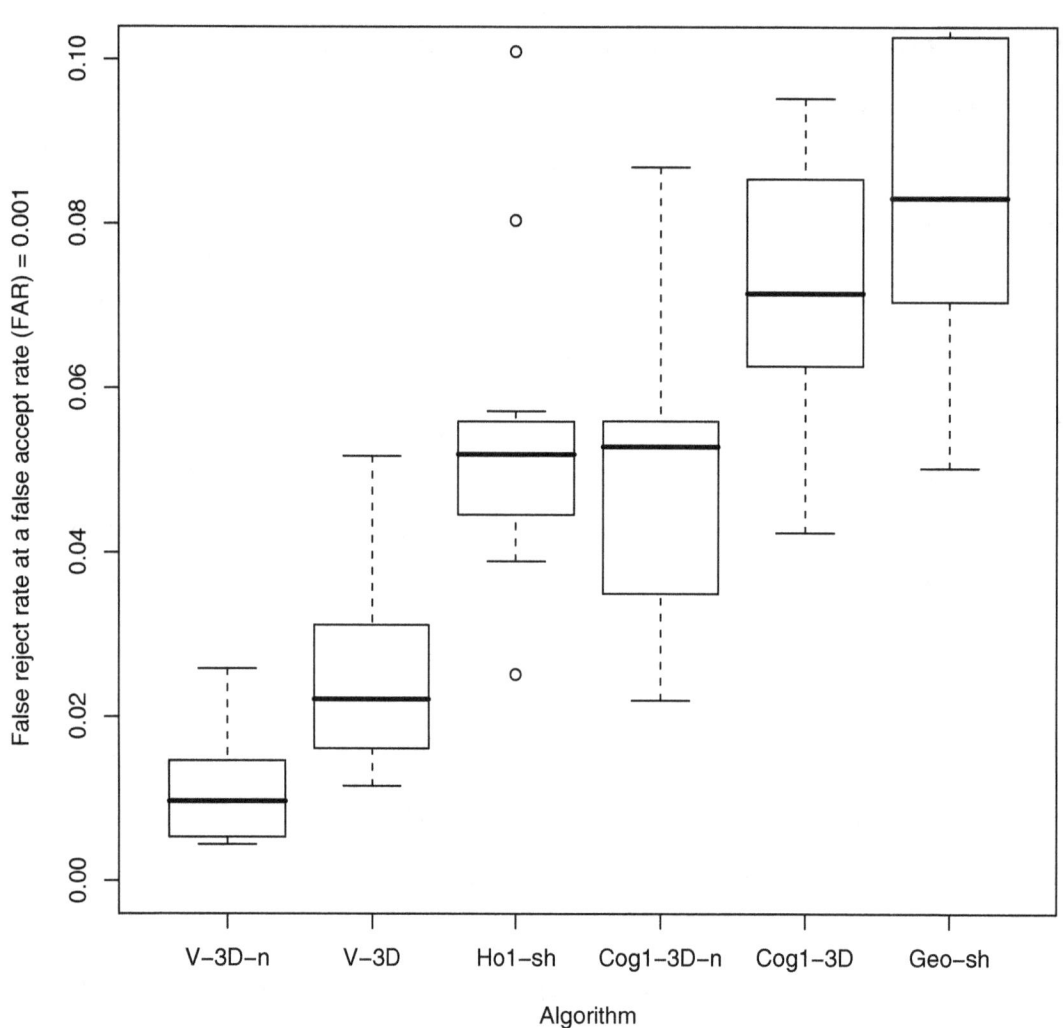

Figure 6: SUMMARY OF PERFORMANCE FOR 3D FACE RECOGNITION ALGORITHMS.

images are frontal facial images taken under *uncontrolled* illumination, see Figure 2 for sample images. These experiments will be referred to as *uncontrolled* experiments.

Performance on controlled versus uncontrolled experiments was measured on the very-high resolution and high-resolution datasets. Figure 7 summarizes the results of the uncontrolled experiments.

In January 2005, the three best self-reported results in the FRGC uncontrolled illumination experiments were FRRs of 0.24, 0.39, and 0.56 at a FAR of 0.001 [6][4]. In FRVT 2006, four algorithms, Cognitec (COG), Neven Vision (NV1-NORM), SAIT (ST-NORM), and Viisage (V-NORM) had performance on both the very-high resolution and high-resolution datasets that was better than the best FRGC results. On the very-high resolution dataset, SAIT (ST-NORM) had a FRR interquartile range of 0.103 to 0.130 at a FAR of 0.001. On the high-resolution dataset Viisage (V-NORM) had a FRR interquartile range of 0.119 to 0.146 at a FAR of 0.001.

In terms of difficulty level, the results in Figure 7 show that there is no clear ranking of the two datasets in terms of difficulty since three algorithms have better performance on the high-resolution dataset; two algorithms had better performance on the very-high resolution datasets; and two algorithms had equivalent performance for both datasets. Restricting our attention to the best results, we see comparable performance for SAIT (ST-NORM) on the very-high resolution dataset and Viisage (V-NORM) on the both datasets.

5.4 Human Performance

FRVT 2006 integrated human face recognition performance into an evaluation for the first time. This inclusion allowed a direct comparison between humans and state-of-the-art computer algorithms. The study focused on recognition across changes in lighting. The experiment matched faces taken under controlled illumination against faces taken under uncontrolled illumination.

Compared with the FRVT 2006 human benchmark, Tsinghua (TS2-NORM) performed better than humans, and Viisage (V-NORM) and SAIT (ST-NORM) were comparable at all

---

[4]These results are on ROC III for Experiment 4 on the FRGC v2 challenge problem.

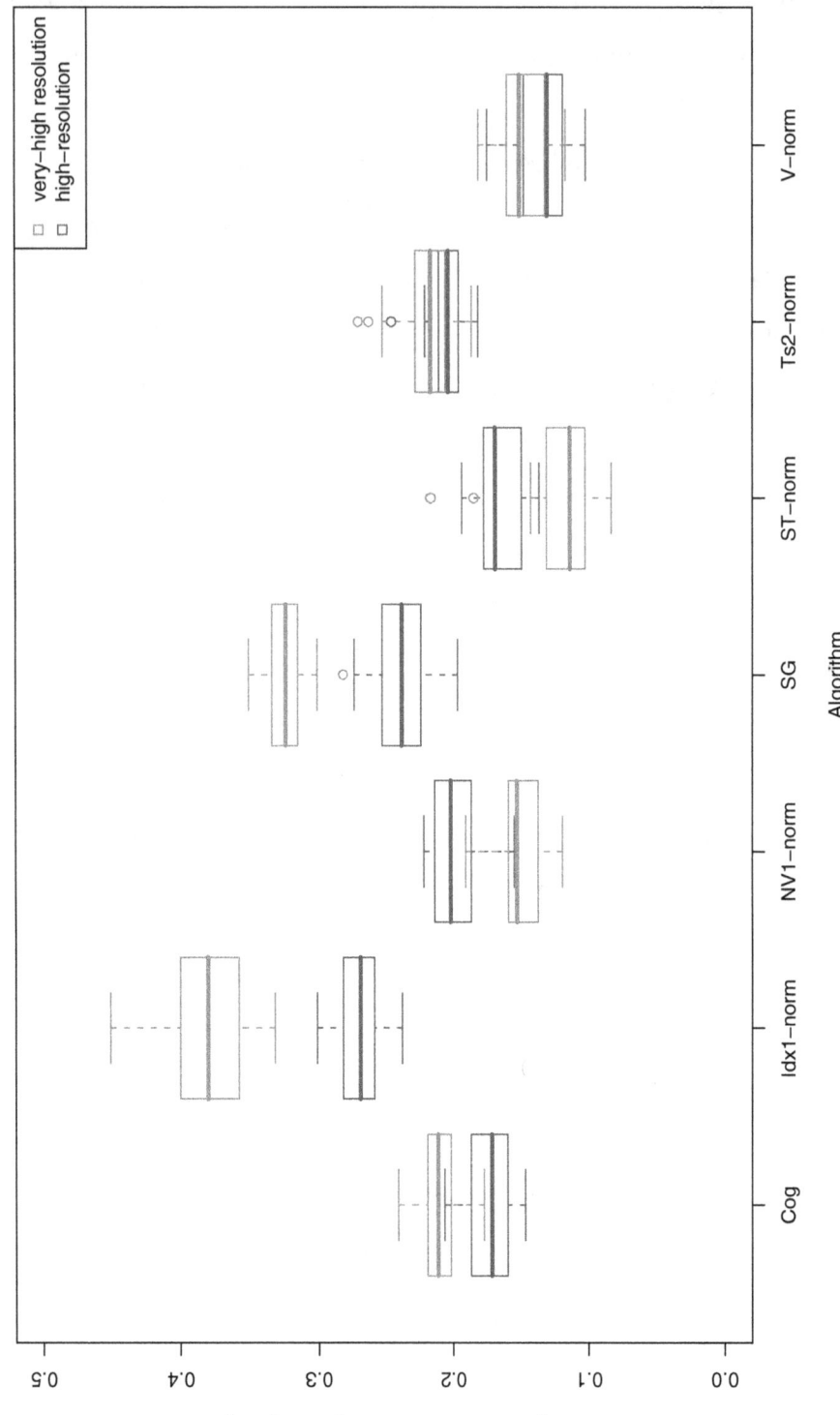

Figure 7: SUMMARY OF STILL FACE RECOGNITION PERFORMANCE ACROSS ILLUMINATION CHANGES ON THE VERY-HIGH RESOLUTION AND HIGH RESOLUTION DATASETS. FOR COGNITEC AND SAGEM, RESULTS FOR THE COG1-NORM AND SG2-NORM ALGORITHMS ARE REPORTED ON THE VERY-HIGH RESOLUTION DATASET, AND RESULTS FOR THE COG1-1TO1 AND SG1-1TO1 ALGORITHMS ARE REPORTED ON THE HIGH-RESOLUTION DATASET.

operating points. Figure 8 compares human and computer performance for the algorithms in Figure 7. Results in Figure 8 are reported on a receiver operating characteristic (ROC) to show the change in relative performance of humans and computers over a range of operating points. Human performance is reported at four operating points (the black dots in Figure 8). The lowest FAR of the four is 0.05. At a FAR of 0.05, six of seven algorithms have the same or better performance than humans. The FRVT 2006 human and machine experiments are in agreement with the results of O'Toole et al. [15] on "difficult" image pairs.

In the human performance experiments, individuals were asked to judge the similarity of 80 pairs of faces. To directly compare performance with face recognition algorithms, performance was computed for seven algorithms for the same 80 face pairs. This experimental design allowed for a direct comparison of humans and algorithms, and followed the design in O'Toole et al. [15]. The only difference is the method for selecting face image pairs.

Since humans can only rate a limited number of pairs of faces, 80 face pairs were selected from the approximately 10 million face pairs that the algorithms compared in the uncontrolled illumination experiments. To gain insight into the relative performance of humans and a set of algorithms, moderately difficult face pairs were selected for this experiment. A face pair is moderately difficult if approximately half of the algorithms performed correctly (e.g., if a face pair were images of the same person, then approximately half of the algorithms reported that the images were of the same person).

The sampling of face pairs was done as follows. All face pairs in the uncontrolled illumination experiment on the high-resolution dataset, see Section 5.3, were given a difficulty score. The difficulty score was based on the number of algorithms that correctly estimated the match status of the face pairs at a FAR of 0.001. For face pairs of the same person, the difficulty score was the number of algorithms that correctly estimated the face pair as the person. Similarly for face pairs of different people, the difficulty score was the number of algorithms that estimated the face pair to be different people. The difficulty score was computed based on the results of eight one-to-one algorithms. The easiest face pairs were assigned the maximum difficulty score of 8 because all eight algorithms assigned the correct match status. The most difficult face pairs were assigned the minimum score of zero, because none of the algorithms assigned the face pair the correct match status. Moderately difficult

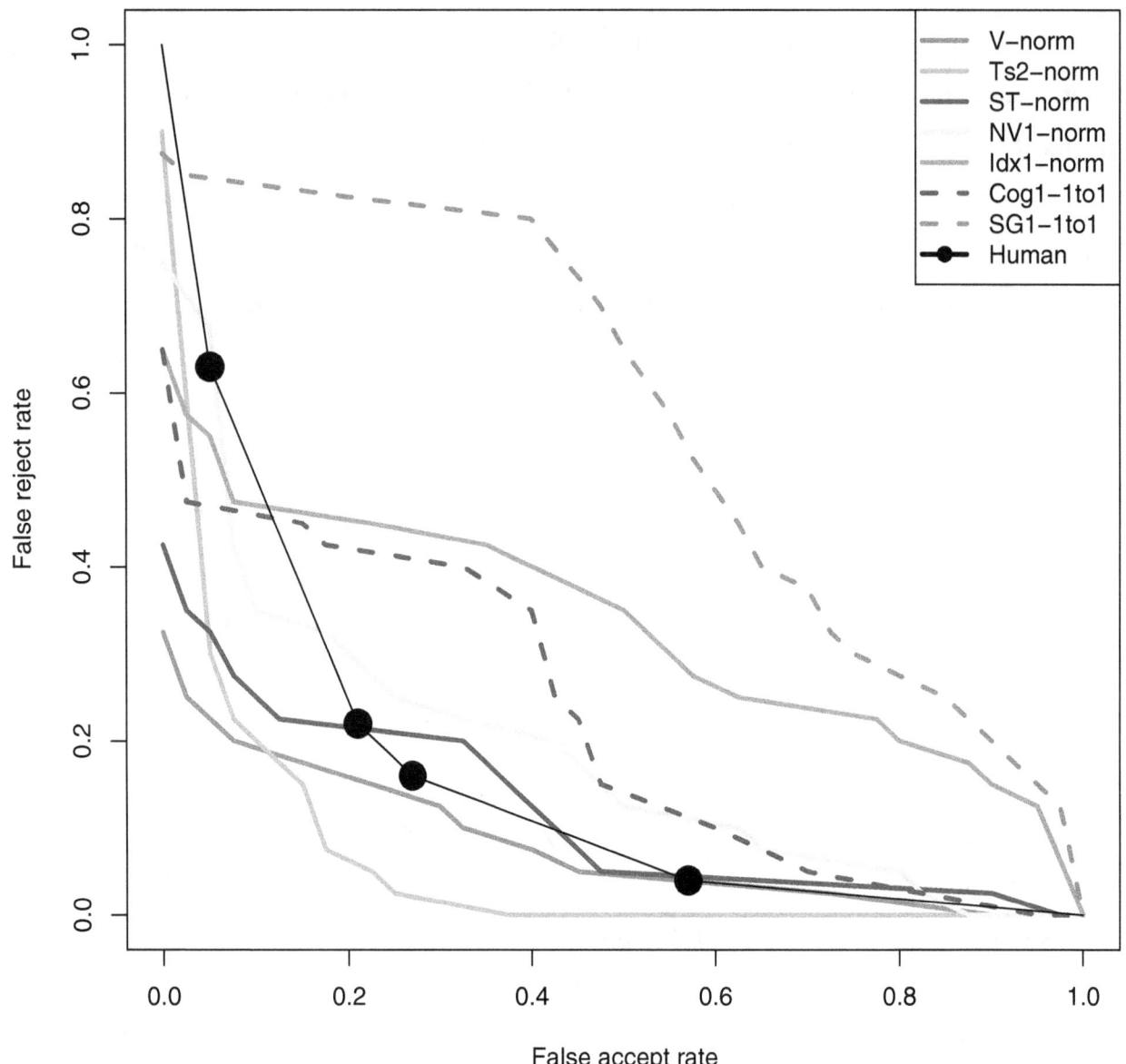

Figure 8: ROC OF HUMAN AND COMPUTER PERFORMANCE ON MATCHING FACES ACROSS ILLUMINATION CHANGES. ROCs FOR ALGORITHMS IN FIGURE 7 ARE PLOTTED. THE ROC PLOTS FAR AGAINST FRR. PERFECT PERFORMANCE WOULD BE THE LOWER LEFT HAND CORNER (FAR=FRR=0).

face pairs with a difficulty of between 3 and 5 were selected for this experiment. From these pairs, we selected 40 pairs of male and 40 pairs of female faces for the human performance experiments. Half of these pairs were match pairs (images of the same person) and half were non-match pairs (images of different people). Face pairs were presented side by side on the computer screen for two seconds. After each pair of faces was presented, subjects rated the similarity of the two faces on a scale of 1 to 5. Subjects responded, using labeled keys on the keyboard as follows: 1.) You are sure they are the same person; 2.) You think they are the same person; 3.) You don't know; 4.) You think they are different people; 5.) You are sure they are different people. A total of 26 undergraduates at the University of Texas at Dallas participated in the experiment.

6 Comparison of Biometric Modalities

FRVT 2006 and ICE 2006 are the first technology evaluations that allowed iris recognition, still face recognition, and 3D face recognition performance to be compared. The results on the multi-biometric dataset show that the performance for all three biometrics is comparable. Figure 9 compares the top performers on each of the three biometrics.

Performance results for the multi-biometric comparison are from the multi-biometric dataset. The multi-biometric dataset is an appropriate dataset for comparing performance across the different biometrics because the dataset controls for population, illumination, and time frame. In this dataset:

- Biometric samples were collected from the same population.

- Biometric samples were collected in the same laboratory during the same time period.

- The samples for all three biometrics were collected under controlled conditions.

    - The iris sensor and 3D sensor have active illumination sources.
    - The still face images were collected under a constant controlled illumination source following the recommendations on the NIST mugshot best practices [16].

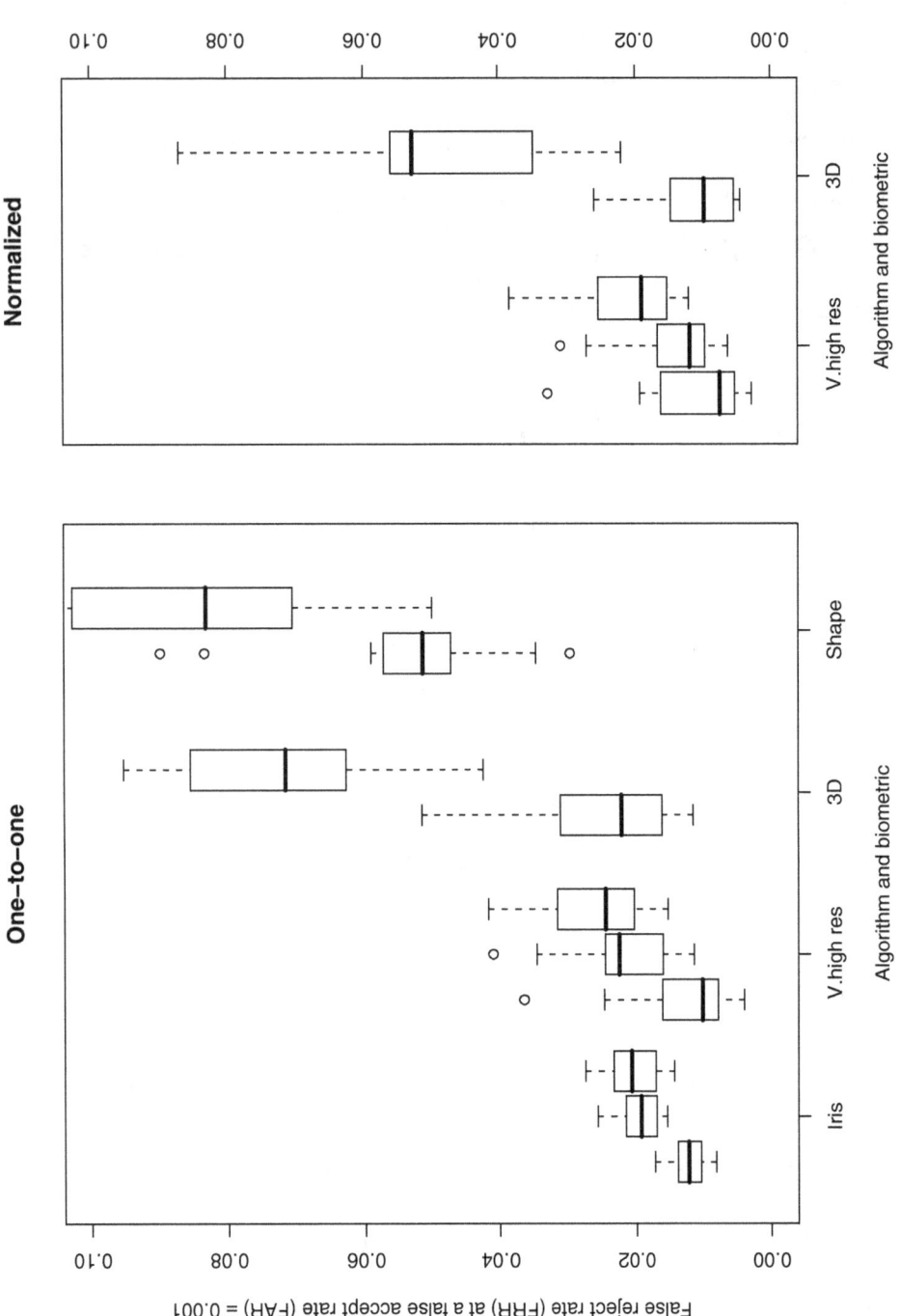

Figure 9: A COMPARISON OF THREE BIOMETRICS: IRIS, HIGH-RESOLUTION STILL FACE, AND 3D FACE. THE LEFT-HAND PANEL REPORTS PERFORMANCE FOR ONE-TO-ONE ALGORITHMS AND THE RIGHT-HAND PANEL REPORTS PERFORMANCE FOR NORMALIZED ALGORITHMS. EACH GROUP ON THE HORIZONTAL AXIS CORRESPONDS TO A BIOMETRIC. FOR EACH BIOMETRIC, THE BEST TWO OR THREE RESULTS ARE PRESENTED. THE RESULTS FOR AN ALGORITHM ARE SUMMARIZED ON A BOXPLOT. THE FALSE REJECT RATE (FRR) AT A FALSE ACCEPT RATE (FAR) OF 0.001 IS REPORTED. THE ALGORITHMS REPORTED ARE SAGEM-IRIDIAN, CAMBRIDGE, AND IRITECH FOR IRIS; NEVEN VISION, VIISAGE, AND COGNITEC FOR STILL FACE; VIISAGE AND COGNITEC FOR 3D FACE; AND HOUSTON AND GEOMETRIX FOR SHAPE. THE RIGHT PANEL REPORTS PERFORMANCE FOR NORMALIZED ALGORITHMS. IN THE RIGHT PANEL, THE ALGORITHMS REPORTED ARE NEVEN VISION, VIISAGE, AND COGNITEC FOR STILL FACE; AND VIISAGE AND COGNITEC FOR 3D FACE.

While the comparison among biometrics in the FRVT 2006 and ICE 2006 evaluation does control for the factors list above, there are other factors that are not controlled. These include maturity of the sensor technology, acquisition time for a biometric sample, cooperation required from a subject, and resolution of the sensor. In general, sensors for 3D biometric imaging of faces are less mature than cameras for iris and face imaging [17]. The 3D sensor used to collect data for the FRVT 2006 has a longer image acquisition time than the iris sensor or digital camera. The iris sensor requires a greater degree of user interaction and cooperation than the 3D sensor; and the 3D sensor requires a greater degree of user interaction and cooperation than the digital camera. Sensors for iris imaging and 3D imaging have fewer sample points than the number of pixels in a normal high-resolution camera image. However, the sensors selected for the multi-biometric dataset collection were representative of the state-of-the-art commercial sensors available at the start of the collection effort.

To be consistent, we compared iris and still face recognition on only one-to-one matching because all the ICE 2006 submissions were one-to-one matching algorithms. The performance of the Sagem-Iridian (SI-2) iris algorithm with a FRR interquartile range of 0.011 to 0.014 at FAR of 0.001 and Neven Vision (NV1-1TO1) still face with a FRR interquartile range of 0.008 to 0.016 at a FAR of 0.001 are comparable.

We compared normalized still and 3D face recognition algorithms because performance with normalized face recognition algorithms was superior to the performance of one-to-one matchers. The performance of the Viisage (V-3D-N) 3D algorithm with a FRR interquartile range of 0.005 to 0.015 at FAR of 0.001 and Neven Vision (NV1-1TO1) still face with a FRR interquartile range of 0.006 to 0.015 at a FAR of 0.001 are comparable.

The results for the Viisage still and 3D submissions show the potential of fusing shape and texture information to improve performance over still imagery alone. For the Viisage still algorithm (V-NORM), the FRR interquartile range was 0.010 to 0.017 at a FAR of 0.001 on the very-high resolution dataset. The Viisage (V-3D-N) 3D algorithm has a FRR interquartile range of 0.005 to 0.015 at FAR of 0.001, where the 3D consists of both shape and texture channels.

To see if the relative performance of face and iris is stable across different false accept rates, we also examined the relative performance at a false accept rate of 0.0001 (one in ten thousand). Considering the number of subjects and biometric samples available, this is the limit of performance that can be measured for face recognition on the multi-modal dataset. At a false accept rate of 0.0001, the relative performance of the NevenVision and iris submissions is the same. The one-to-one Cognitec and one-to-one Viisage submissions are not comparable with the iris submissions. However, the performance of their normalization submissions is comparable to the one-to-one iris submissions.

7 Discussion and Conclusion

Considered together, the FRVT 2006 and the ICE 2006 form the first multi-biometric technology evaluation designed to compare performance across biometrics. This evaluation established the first performance benchmark for iris and 3D face recognition, assessed the advancement in the state-of-the-art in still face recognition, and compared performance across the biometrics evaluated. Figures 10 and 11 summarize the results of the FRVT 2006 and the ICE 2006 for all three biometrics and all datasets.

Face recognition performance on still frontal images taken under controlled illumination has improved by an order of magnitude since the FRVT 2002. There are three primary components to the improvement in algorithm performance since the FRVT 2002: a) the recognition technology, b) higher resolution imagery, and c) improved quality due to greater consistency of lighting. Since performance was measured on the low-resolution dataset in both the FRVT 2002 and the FRVT 2006, it is possible to estimate the improvement in performance due to algorithm design. The improvement in algorithm design resulted in an increase in performance by a factor of between four and six depending on the algorithm. For the results on the high and very-high resolution datasets, the improvement in performance comes from a combination of algorithm design and image size and quality. This is because new recognition techniques have been developed to take advantage of the larger high quality face images. The performance on the very-high resolution dataset shows one path for improving the performance of face recognition systems. The existence of the very-high resolution dataset shows high quality data can be collected in large scale laboratory collection efforts. One of the challenges for the face recognition community is to develop acquisition

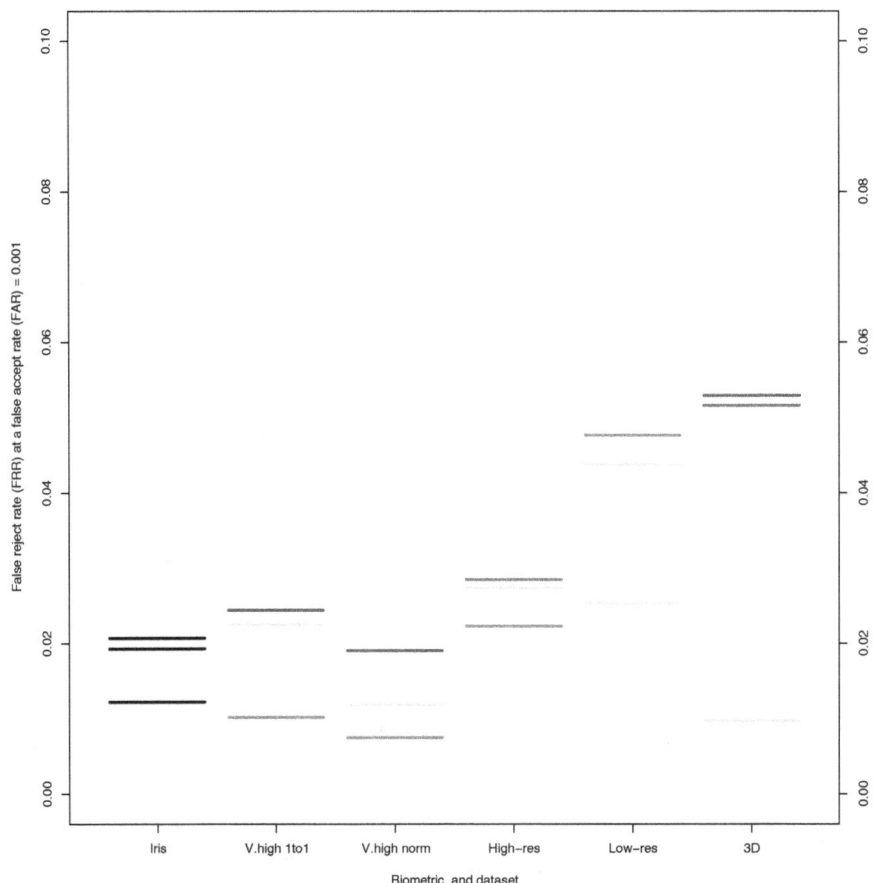

Figure 10: SUMMARY OF PERFORMANCE ON FRVT 2006 AND ICE 2006 FOR CONTROLLED ILLUMINATION EXPERIMENTS. THE FIRST COLUMN (LABELED IRIS) PLOTS THE MEDIAN FRR AT A FAR OF 0.001 FOR THE THREE IRIS ALGORITHMS IN FIGURE 3. (ALL PERFORMANCE SCORES REPORTED IN THIS GRAPH ARE FRR AT A FAR OF 0.001.) THE SECOND COLUMN (V.HIGH 1TO1) REPORTS FRR FOR THE TOP THREE ONE-TO-ONE STILL FACE RECOGNITION ALGORITHMS ON VERY-HIGH RESOLUTION DATASET. THE THIRD COLUMN (V.HIGH NORM) REPORTS FRR FOR THE TOP THREE NORMALIZED STILL FACE RECOGNITION ALGORITHMS ON THE VERY-HIGH RESOLUTION DATASET; THE FORTH (HIGH-RES) AND FIFTH (LOW-RES) REPORT FRR FOR NORMALIZED ALGORITHMS ON THE HIGH-RESOLUTION AND LOW-RESOLUTION DATASETS; AND THE SIXTH COLUMN (3D) REPORTS FRR FOR NORMALIZED 3D FACE RECOGNITION ALGORITHMS. FOR THE FACE RECOGNITION RESULTS, ALL ALGORITHMS FROM THE SAME GROUP ARE THE SAME COLOR.

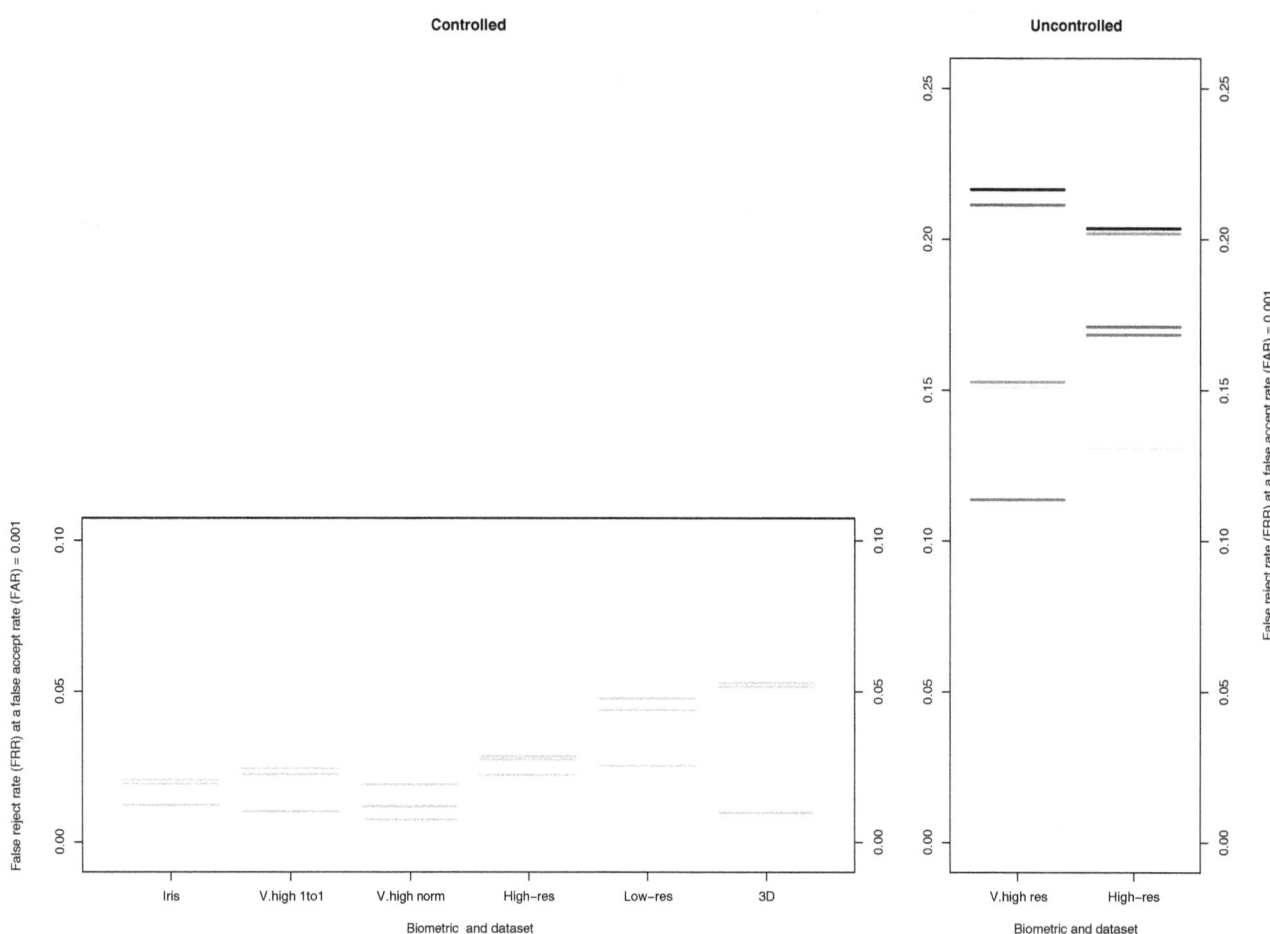

Figure 11: SUMMARY OF PERFORMANCE ON FRVT 2006 AND ICE 2006. THE LEFT PANEL IS FIGURE 10 RESCALED TO ALLOW FOR COMPARISONS BETWEEN RESULTS ON THE CONTROLLED AND UNCONTROLLED ILLUMINATION EXPERIMENTS. THE RIGHT PANEL SUMMARIZES RESULTS FOR THE UNCONTROLLED ILLUMINATION EXPERIMENTS. ALL PERFORMANCE SCORES REPORTED IN THIS GRAPH ARE FRR AT A FAR OF 0.001. IN THE RIGHT PANEL, THE FIRST COLUMN (V.HIGH RES) PLOTS THE MEDIAN FRR FOR FIVE NORMALIZED FACE RECOGNITION ALGORITHMS ON VERY-HIGH RESOLUTION DATASET. THE SECOND COLUMN (HIGH-RES) REPORTS FRR FOR FIVE NORMALIZED ALGORITHMS ON THE HIGH-RESOLUTION DATASET. IN THE UNCONTROLLED ILLUMINATION PANEL, RESULTS FOR ALGORITHMS FROM THE SAME GROUP ARE THE SAME COLOR.

techniques, protocols, and systems that allow for this quality of data to be collected in fielded applications.

The ability of algorithms to recognize faces across illumination changes has made significant progress. The FRVT 2006 measured progress on this problem by matching images taken under uncontrolled illumination against images taken under controlled illumination. In the FRVT 2006, the performance of matching controlled images against uncontrolled images is at a FRR of less than 0.20 at FAR = 0.001. In FRVT 2002, the corresponding performance for matching controlled images against controlled images was a FRR of 0.20 at FAR = 0.001. Thus, performance from uncontrolled images in FRVT 2006 is better than performance from controlled images in FRVT 2002.

The difference between the design of the controlled and uncontrolled illumination experiments was the probe images. In both experiments, the same set of controlled illumination images was used for the enrolled images. In the controlled experiments, the probe images were also taken under the same controlled light conditions; in the uncontrolled experiments, the probe images were taken under uncontrolled illumination conditions. The FRVT 2006 results show that relaxing the illumination condition has a dramatic effect on performance, see Figure 11. For the controlled illumination experiments, performance of the very-high resolution dataset was better than the high-resolution dataset. By contrast, relaxing the illumination constraints on the probe images resulted in comparable performance on the very-high resolution and high resolution datasets.

The human visual system contains a very robust face recognition capability that is excellent at recognizing familiar faces [18]. However, human face recognition capabilities on unfamiliar faces falls far short of the capability for recognizing familiar faces. The FRVT 2006, for the first time, integrated measuring human face recognition capability into an evaluation. Performance of humans and computers was compared on the same set of images. The FRVT 2006 human and computer experiment measures the ability to recognize faces across illumination changes. This experiment found that algorithms are capable of human performance levels, and that at false accept rates in the range of 0.05, machines can out-perform humans.

The multi-biometric component of ICE 2006 and FRVT 2006 allowed for a direct comparison of iris, still face, and 3D face recognition technology. One of the findings of this report is that on the multi-biometric dataset, performance of all three biometrics is comparable when all three biometrics are acquired under controlled illumination. For iris and 3D face, the sensor contains an active illumination source and for still face the data was collected under static controlled lighting. The comparison between iris and still face is summarized in Figure 10 in the columns labeled *iris* and *V.high 1to1*. The corresponding comparison is shown in columns *V.high norm* and *3D*.

FRVT 2006 is the sixth in a series of U.S. Government sponsored face recognition technology evaluations. In thirteen years, performance has improved by two orders of magnitude and there exist numerous companies selling face recognition systems. The evaluations provided regular assessments of the state of the technology and helped to identify the most promising approaches. The challenge problems also nurtured research efforts by providing large datasets for use in developing new algorithms. The FERET, FRGC and FRVT evaluations and challenge problems were instrumental in advancing face recognition technology, and they show the potential for the evaluation and challenge problem paradigm to advance biometric, pattern recognition, and computer vision technologies.

References

[1] P. J. Phillips, A. Martin, C. L. Wilson, and M. Przybocki, "An introduction to evaluating biometric systems," *Computer*, vol. 33, pp. 56–63, 2000.

[2] W. Zhao, R. Chellappa, P. J. Phillips, and A. Rosenfeld, "Face recognition: A literature survey," *ACM Computer Surveys*, vol. 35, pp. 399–458, 2003.

[3] L. Flom and A. Safir, "Iris recognition system," U.S. Patent 4,641,349, 1987.

[4] K. W. Bowyer, K. Hollingsworth, and P. J. Flynn, "Image understanding for iris biometrics: A survey," Department of Computer Science and Engineering, U of Notre Dame, Tech. Rep., 2007.

[5] P. J. Phillips, P. J. Flynn, T. Scruggs, K. W. Bowyer, J. Chang, K. Hoffman, J. Marques, J. Min, and W. Worek, "Overview of the face recognition grand challenge," in *IEEE*

Computer Society Conference on Computer Vision and Pattern Recognition, 2005, pp. 947–954.

[6] P. J. Phillips, P. J. Flynn, W. T. Scruggs, K. W. Bowyer, and W. Worek, "Preliminary face recognition grand challenge results," in *Seventh International Conference on Automatic Face and Gesture Recognition*, 2006, pp. 15–24.

[7] P. Phillips, P. Grother, R. Micheals, D. Blackburn, E. Tabassi, and J. Bone, "Face recognition vendor test 2002: Evaluation report," National Institute of Standards and Technology, Tech. Rep. NISTIR 6965, 2003, http://www.frvt.org.

[8] J. Daugman, "High confidence visual recognition of persons by a test of statistical independence," *IEEE Trans. PAMI*, vol. 15, no. 11, pp. 1148–1161, 1993.

[9] International Biometric Group, "Independent testing of iris recognition technology," International Biometric Group, Tech. Rep., May 2005. [Online]. Available: http://www.ibgweb.com/reports/public/ITIRT.html

[10] M. Turk and A. Pentland, "Eigenfaces for recognition," *J. Cognitive Neuroscience*, vol. 3, no. 1, pp. 71–86, 1991.

[11] H. Moon and P. J. Phillips, "Computational and performance aspects of PCA-based face-recognition algorithms," *Perception*, vol. 30, pp. 303–321, 2001.

[12] P. J. Phillips, H. Moon, S. Rizvi, and P. Rauss, "The FERET evaluation methodology for face-recognition algorithms," *IEEE Trans. PAMI*, vol. 22, pp. 1090–1104, October 2000.

[13] L. Wiskott, J.-M. Fellous, N. Kruger, and C. von der Malsburg, "Face recognition by elastic bunch graph matching," *IEEE Trans. PAMI*, vol. 17, no. 7, pp. 775–779, 1997.

[14] K. Okada, J. Steffens, T. Maurer, H. Hong, E. Elagin, H. Neven, and C. von der Malsburg, "The Bochum/USC face recognition system," in *Face Recognition: From Theory to Applications*, H. Wechsler, P. J. Phillips, V. Bruce, F. Fogelman Soulie, and T. S. Huang, Eds. Berlin: Springer-Verlag, 1998, pp. 186–205.

[15] A. J. O'Toole, P. J. Phillips, F. Jiang, J. Ayyad, N. Pénard, and H. Abdi, "Face recognition algorithms surpass humans matching faces across changes in illumination," *IEEE Trans. PAMI*, in press 2007.

[16] R. M. McCabe, "Best practice recommendation for the capture of mugshots version 2.0," 1997, http://www.nist.gov/itl/div894/894.03/face/face.html.

[17] K. W. Bowyer, K. Chang, and P. J. Flynn, "A survey of approaches and challenges in 3D and multi-modal 3D+2D face recognition," *Computer Vision and Image Understanding*, vol. 101, no. 1, pp. 1–15, January 2006.

[18] P. J. B. Hancock, V. Bruce, and A. M. Burton, "Recognition of unfamiliar faces," *Trends in Cognitive Sciences*, vol. 4, pp. 330–337, 2000.

A-1 Materials and Methods

A-1.1 Data

Data for the FRVT 2006 and ICE 2006 came from three sources: the University of Notre Dame, Sandia National Laboratories, and the U.S. Government (which provided the HCInt data set from FRVT 2002). The data collected at Notre Dame was used in both FRVT 2006 and ICE 2006, while the Sandia data was collected specifically for FRVT 2006. All data used in the FRVT 2006 and ICE 2006 evaluations was sequestered at the subject level, meaning the data was not released to the public and subjects in the FRVT 2006 and ICE 2006 datasets were not included in any previous evaluations or challenge problems; e.g., FRGC and ICE 2005.

The still facial image data collected at Notre Dame and Sandia allowed performance to be measured on very-high and high resolution still images. The data collected at University of Notre Dame is part of an ongoing multi-biometric data collection. The University of Notre Dame collected multi-modal data consisting of high resolution face still, 3D face and iris images. Data for FRVT 2006 was collected during the Fall 2004 and Spring 2005 semesters and data for ICE 2006 was collected during the Spring 2004, Fall 2004, and Spring 2005 semesters.

Data at Notre Dame was collected in subject sessions. A *subject session* is the set of all images (still face, 3D face, and iris) of a person taken each time a person's biometric data is collected. Subjects were invited to participate in acquistion sessions at roughly weekly intervals throughout the academic year. As a result, there are many multi-biometric subject sessions in the dataset, which allows for a comparison among different biometrics.

The FRVT 2006 data for a subject session consists of two controlled still images, two uncontrolled still images, and one three-dimensional image. Figure 2 shows a set of images for one subject session. The controlled images were taken in a studio setting and are full frontal facial images taken with two facial expressions (neutral and smiling). The uncontrolled images were taken in varying illumination conditions; e.g., hallways, atria, or outdoors. Each set of uncontrolled images contains two expressions: neutral and smiling.

The still images were taken with a 6 Megapixel Nikon D70 digital SLR camera with a stock 18-70mm lens. For the controlled images, the camera was configured to maximize the face size in the image. The camera was configured to automatically focus and adjust exposure times for an indoor picture. The controlled still images are 3008 pixels high by 2000 pixels wide; the uncontrolled still images are 2000 pixels high by 3008 pixels wide. Images were saved in a proprietary Nikon exchange format (NEF) format, converted losslessly to 12-bit TIFF format and converted to JPEG.

The 3D images were acquired by a Minolta Vivid 900/910 series sensor. The Minolta Vivid 900/910 series is a structured light sensor that takes a 640 by 480 range sampling and a registered color image. Subjects stood or sat approximately 1.5 meters from the sensor. The images for the FRVT 2006 were acquired using the 900/910's full resolution "fine" scanning mode. The 3D images were taken indoors under illumination conditions that were appropriate for the Vivid 900/910 sensor but not the same controlled conditions as used for the controlled still images. (The Minolta sensor cannot tolerate high levels of ambient light such as direct sunlight or bright studio lighting used for collecting the very-high resolution images.) These 3D images consist of both range and texture channels. The range channel is acquired during a scanning period of about two seconds in duration; the color images are subsequently acquired from a color wheel that is rotated in front of the scanner's camera. Subject motion during 3D scanning can distort the shape data; motion between 3D scanning and color image acquisition can deregister the shape and texture channels; and motion during the color acquisition can deregister the individual color planes of the texture image.

The ICE 2006 images were acquired using an LG EOU 2200 iris scanner. The LG EOU 2200 is a complete acquisition system and has automatic image quality control checks. By agreement between U. of Notre Dame and Iridian, a modified version of the acquisition software was provided. The modified software allowed all images from the sensor to be saved under certain conditions, as explained below.

The iris images are 480x640 in resolution, see Figure 2. For most "good" iris images, the diameter of the iris in the image exceeds 200 pixels. The images are stored with 8 bits of intensity, but every third intensity level is unused. This is the result of a contrast stretching automatically applied within the LG EOU 2200 system.

In our acquisitions, the subject was seated in front of the system. The system provides recorded voice prompts to aid the subject to position their eye at the appropriate distance from the sensor. The system takes images in "shots" of three, with each image corresponding to illumination of one of the three infrared (IR) light emitting diodes (LED)s used to illuminate the iris.

For a given subject at a given iris acquistion session, two "shots" of three images each are taken for each eye, for a total of 12 images. The system provides a feedback sound when an acceptable shot of images is taken. An acceptable shot has one or more images that pass the LG EOU 2200's built-in quality checks, but all three images are saved. If none of the three images pass the built-in quality checks, then none of the three images are saved. At least one third of the iris images do pass the Iridian quality control checks, and up to two thirds do not pass.

A manual quality control step at Notre Dame was performed to remove images in which, for example, the eye was not visible at all due to the subject having turned their head.

The high-resolution dataset was collected at Sandia National Laboratories over a period of a few years. The high-resolution collection protocol followed the Notre Dame protocol used to collect the very-high resolution dataset. The still images were taken with a 4 Megapixel Canon PowerShot G2. The high-resolution dataset consisted of controlled images taken in a studio setting and uncontrolled images taken in hallways, conference rooms and outside.

The low-resolution dataset allowed for the assessment of performance on a large data set and a direct comparison with performance in FRVT 2002. The low-resolution dataset was used in the FRVT 2002 for the High Computational Intensity (HCInt) test [7].

A-1.2 FRVT 2006 and ICE 2006 Test Protocol

Both the FRVT 2006 and the ICE 2006 were algorithm evaluations in which participants had to deliver algorithms to NIST to be evaluated. For the large scale portion of the FRVT 2006, both one-to-one matching and normalization techniques were evaluated, and for the ICE 2006 one-to-one matching techniques were evaluated. The FRVT 2006 and the ICE 2006 were open to academia, industry, and research laboratories.

The FRVT 2006 and the ICE 2006 used a protocol structure that is substantially different from previous evaluations. The format for submissions was binary executables that could be run independently on the test server. All submitted executables had to run using a specified set of command line arguments. The command line arguments included an experiment parameter file, files that contained the sets of biometric samples to be matched, and name of the output similarity file. An image quality task was also available. Participants could submit multiple algorithms.

The test system hardware for the FRVT 2006 and the ICE 2006 was a Dell PowerEdge 850 server with a single Intel Pentium 4 3.6GHz 660 processor and 2MB of 800Mhz cache. All systems had 4GB of 533MHz DDR2 Ram. At no time did the test system have access to the Internet. The FRVT 2006 and the ICE 2006 allowed executables that would run under Windows Server 2003 (standard edition) and Linux Fedora Core 3 operating systems.

A-1.3 Experimental Design

Performance on the HCInt dataset was computed on a set of *twelve small galleries* generated from the large HCInt gallery. Each gallery consisted of 3,000 individuals with one image per subject. A gallery is a set of biometric samples provided to an algorithm that represents the set of enrolled people. A gallery contains only one biometric sample per person. The twelve small galleries are disjoint (a subject in the HCInt dataset was in only one gallery). There were twelve corresponding small probe sets which consisted of 12,000 images each. A probe is a biometric sample that is presented to an algorithm for verification. Two images of each individual in the corresponding gallery were placed in the probe set and two images of 3,000 individuals not in the gallery were placed in the probe set. The construction of the probe set made it possible to compute *true impostor* verification performance. A *true impostor* is a biometric sample where the subject in the sample is not in the gallery. Table 2 list the number of images, subjects, and partitions for each FRVT 2006 and ICE 2006 experiments.

One-to-one algorithms were required to compute a complete similarity matrix of similarity scores between all biometric samples in a target and probe set. A target set is similar to a gallery with one difference. A target set can contain multiple biometric samples per person. In FRVT 2006 and ICE 2006, the target and probe sets contain multiple biometric samples per subject. One of the first steps in evaluating normalized algorithm is for the test administer to create a set of galleries with one sample per subject. In FRVT 2006, the target

Table 2: Summary of experiments on multi-biometric and high-resolution datasets.

Experiment	Dataset	No. subjects	No. images	No. Partitions
Controlled-face	very-high resolution	336	7496	26
Controlled-face	high-resolution	263	14,365	20
Controlled-face	low-resolution	36,000	108,000	12
Uncontrolled-face	very-high resolution	335	5402	26
Uncontrolled-face	high-resolution	257	7192	20
3D-face	3D	330	3589	13
Iris right-eye	iris	240	29,056	30
Iris left-eye	iris	240	30,502	30

set was divided into a set of n non-overlapping galleries. Performance is then computed for each of the n galleries and the set of n results is then reported on a boxplot. Normalized algorithms were required to compute n similarity matrices, one for each of the n galleries.

A-2 Detailed Results

The results presented in body of this report are a subset of all large-scale experiments that were run for the FRVT 2006 and the ICE 2006. This appendix contains detailed results for all large-scale experiments run for the FRVT2006 and the ICE2006. The graphs include results on the very-high resolution, high-resolution, and low-resolution, 3D, and iris datasets. Results are presented for both one-to-one and normalized algorithms. Results for algorithms that did not complete the large-scale experiments are not shown in this report.

Table 3 list the algorithms whose results are reported in this appendix, and includes the abbreviations used for algorithms in the graphs. Table 4 lists experiments reported, specifying illumination condition, biometric, dataset, whether the algorithm was one-to-one or normalized, and the corresponding figure number for each graph. Following the pattern in the main report, performance is reported on boxplots. In this appendix, performance is reported at three FARs: 0.01, 0.001, and 0.0001. For each algorithm, a boxplot is shown for each of the three FARs.

Table 3: THE LIST OF ALGORITHMS COVERED IN THE LARGE SCALE ANALYSIS. THE ORGANIZATION THAT SUBMITTED AN ALGORITHM IS LISTED IN THE GROUP COLUMN. THE ABBREVIATIONS USED IN THE FIGURES ARE PRESENTED IN THE TABLE.

Group	Iris	Still 1to1	Still norm	3D 1to1	3D norm	Shape
U. of Cambridge	Cam-1					
	Cam-2					
	Cam-3					
	Cam-4					
	Cam-5					
	Cam-6					
Cognitec		Cog1-1to1	Cog1-norm	Cog1-3D	Cog1-3D-n	
		Cog2-1to1	Cog2-norm			
Geometrix						Geo-Sh
U. of Houston						Ho1-Sh
						Ho3-Sh
Identix		Idx4-1to1	Idx1-norm			
		Idx5-1to1	Idx2-norm			
			Idx3-norm			
Iritech	Irtch-1					
	Irtch-2					
Neven Vision		NV1-1to1	NV1-norm			
		NV2-1to1	NV2-norm			
Panvista		Pn-1to1		Pn-3D		
Rafael		Ra-1to1	Ra-norm			
Sagem		SG1-1to1	SG1-norm			
		SG2-1to1	SG2-norm			
Sagem-Iridian	SI-1					
	SI-2					
SAIT		ST-1to1	ST-norm			
Tili		Ti-1to1				
Toshiba		To1-1to1	To1-norm			
		To2-1to1				
Tsinghua U.		Ts-1to1	Ts1-norm	Ts1-3D		
		Ts2-1to1	Ts2-norm			
Viisage		V-1to1	V-norm	V-3D	V-3D-n	
		Va-1to1				

Table 4: List of graphs in Appendix A-2.

Illumination	Biometric	Dataset	Executable class	Full FRR range	Zoomed FRR range
controlled	still-face	very-high resolution	one-to-one	12	13
controlled	still-face	very-high resolution	norm	14	15
controlled	still-face	high-resolution	one-to-one	16	17
controlled	still-face	high-resolution	norm	18	19
controlled	still-face	low-resolution	norm	20	
controlled	3D-face	3D	one-to-one	21	22
controlled	3D-face	3D	norm		23
uncontrolled	still-face	very-high resolution	one-to-one	24	
uncontrolled	still-face	very-high resolution	norm	25	
uncontrolled	still-face	high-resolution	one-to-one	26	
uncontrolled	still-face	high-resolution	norm	27	
controlled	iris-left	iris	one-to-one		28
controlled	iris-right	iris	one-to-one		29

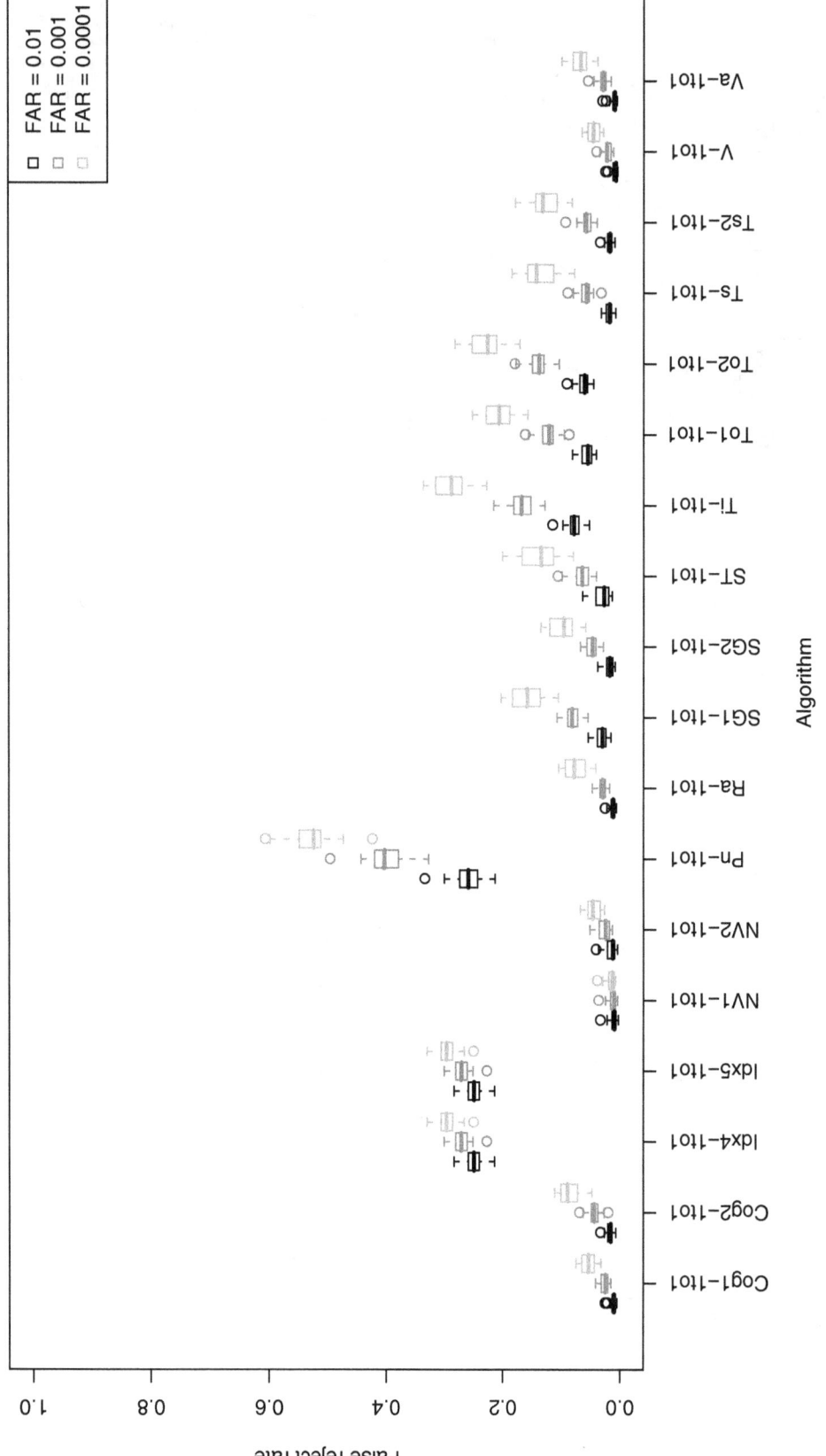

Figure 12: Results for controlled experiment on the very-high resolution dataset for one-to-one algorithms.

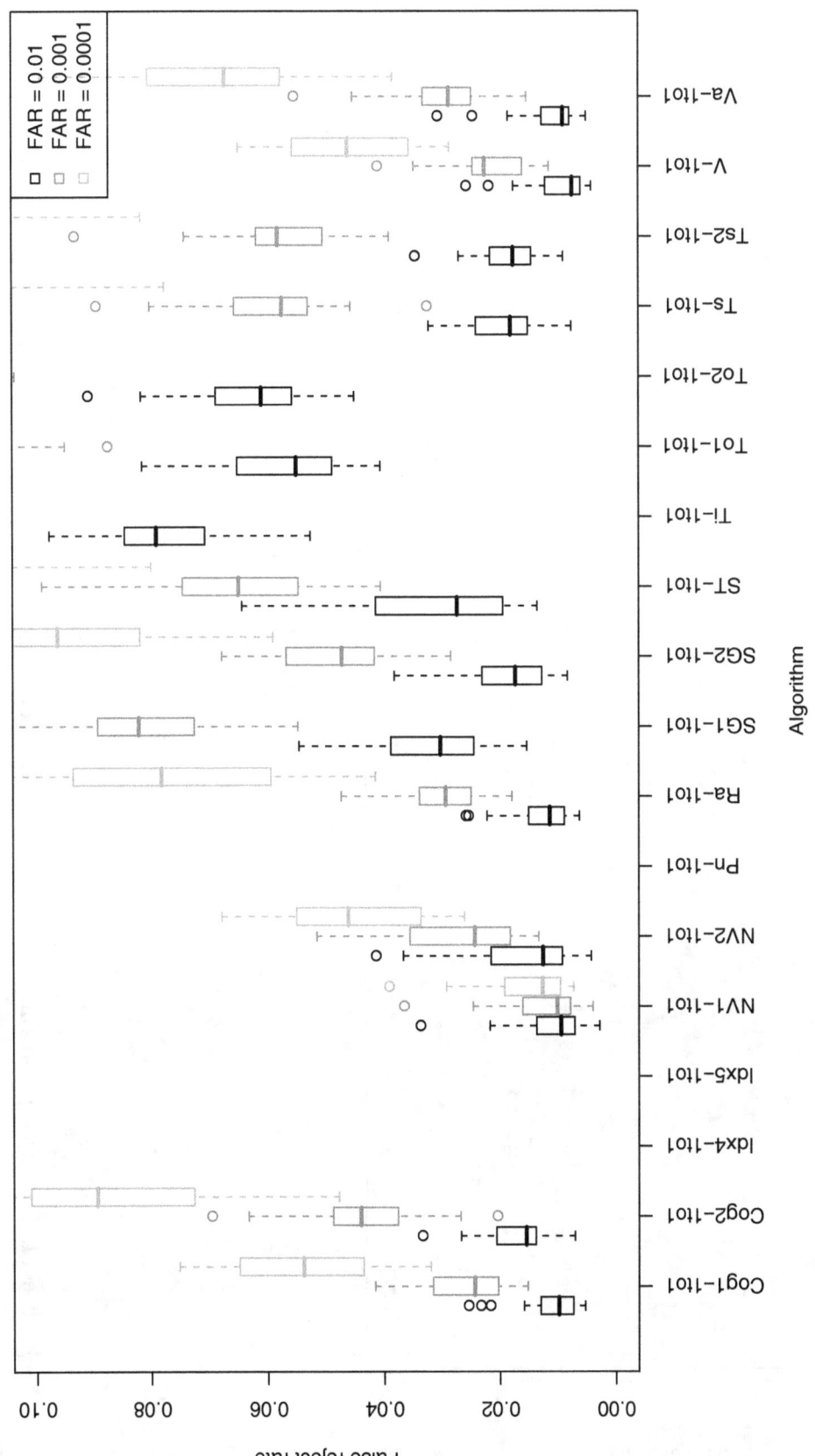

Figure 13: Results for controlled experiment on the very-high resolution dataset for one-to-one algorithms. The range for FRR on the vertical axis is 0.00 to 0.10.

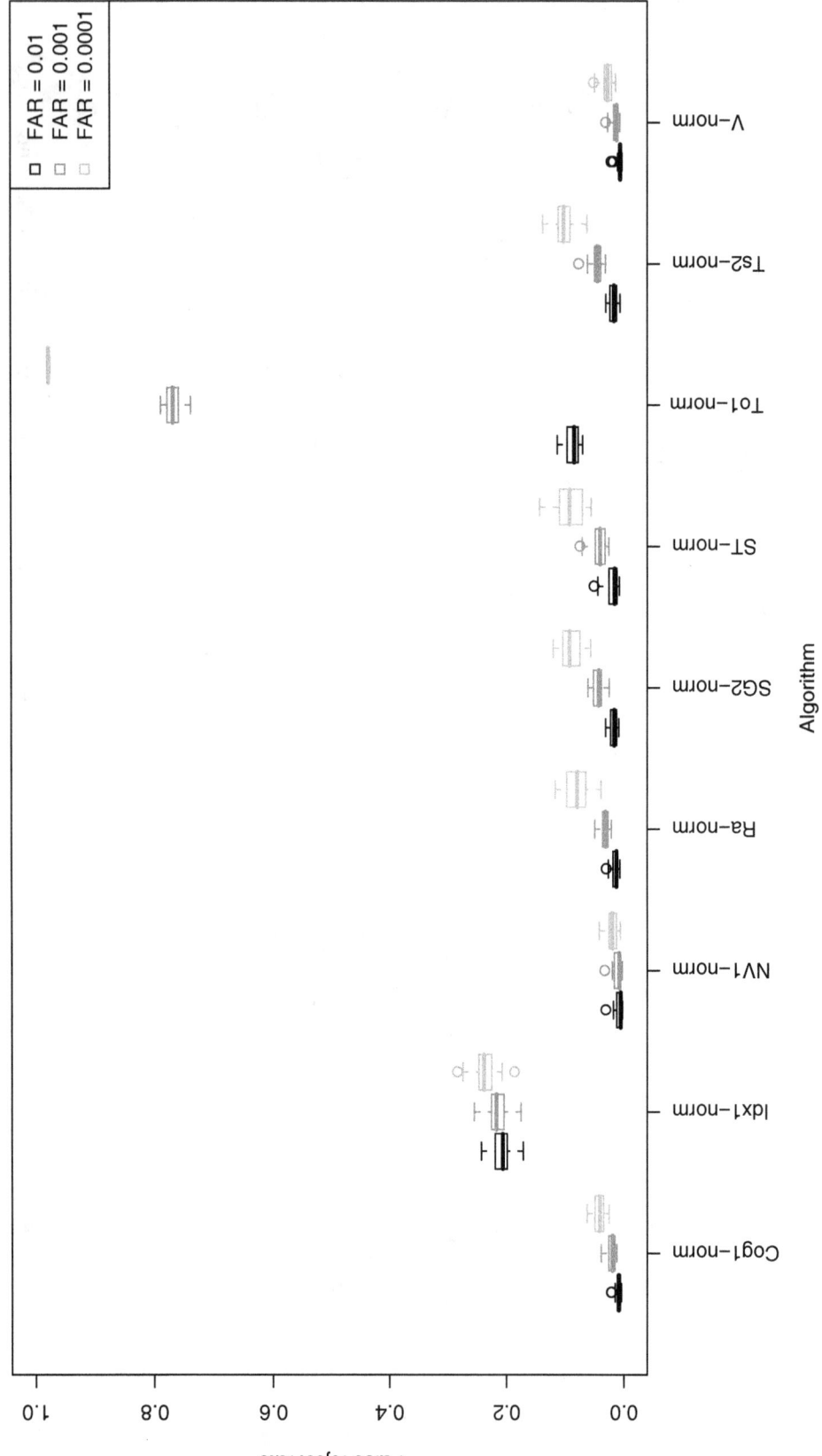

Figure 14: Results for controlled experiment on the very-high resolution dataset for normalized algorithms.

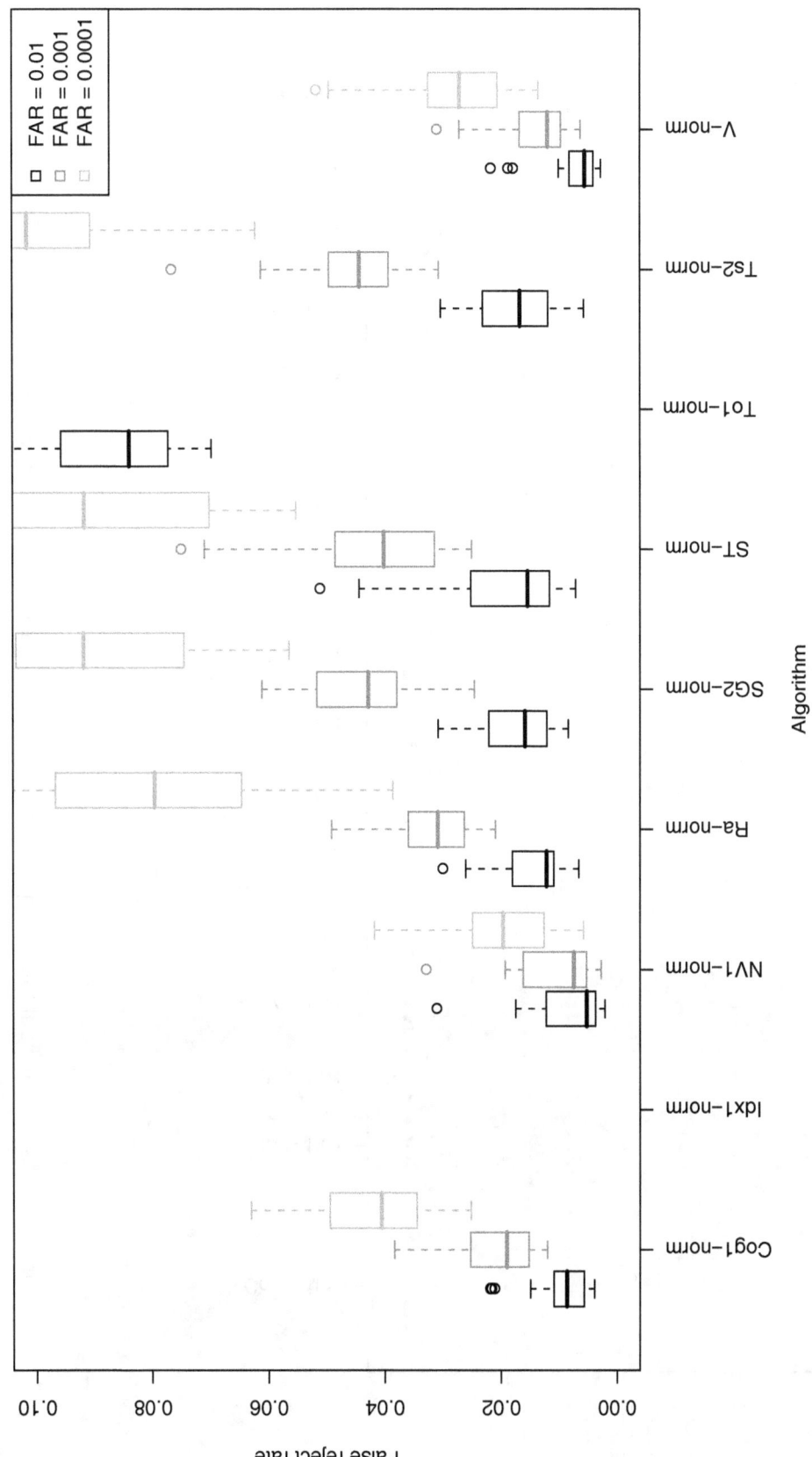

Figure 15: Results for controlled experiment on the very-high resolution dataset for normalized algorithms. The range for FRR on the vertical axis is 0.00 to 0.10.

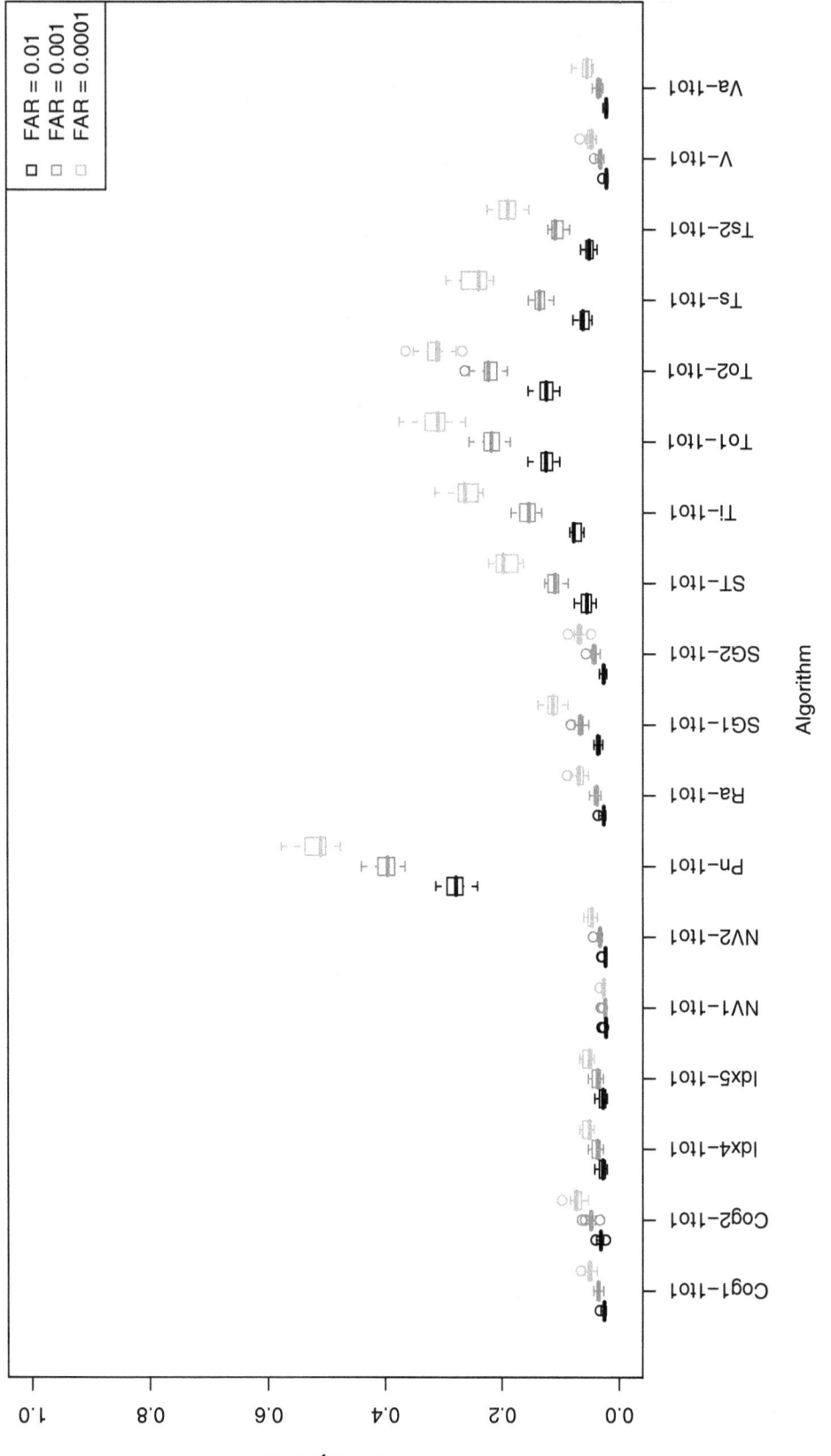

Figure 16: Results for controlled experiment on the high-resolution dataset for one-to-one algorithms.

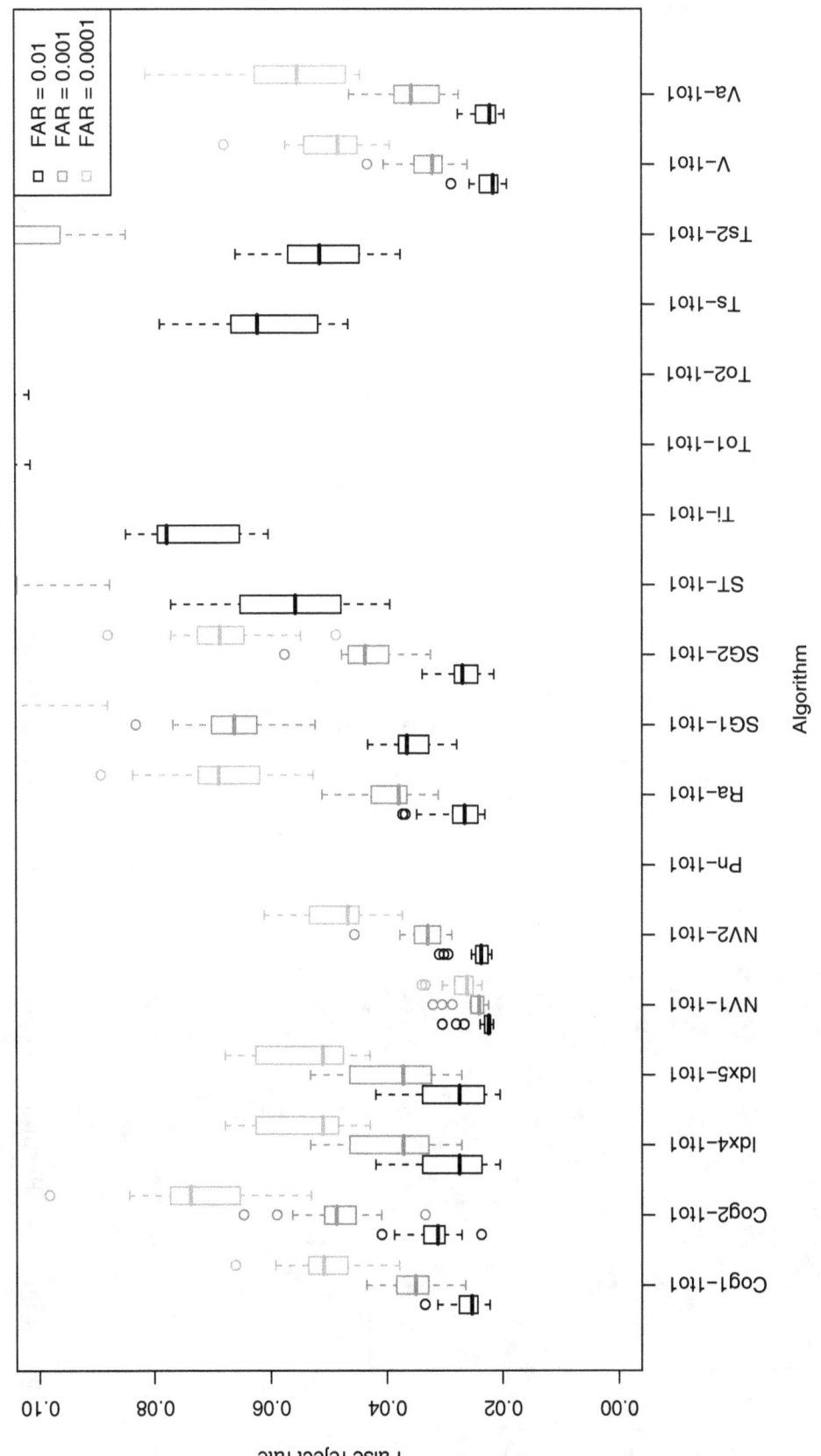

Figure 17: Results for controlled experiment on the high-resolution dataset for one-to-one algorithms. The range for FRR on the vertical axis is 0.00 to 0.10.

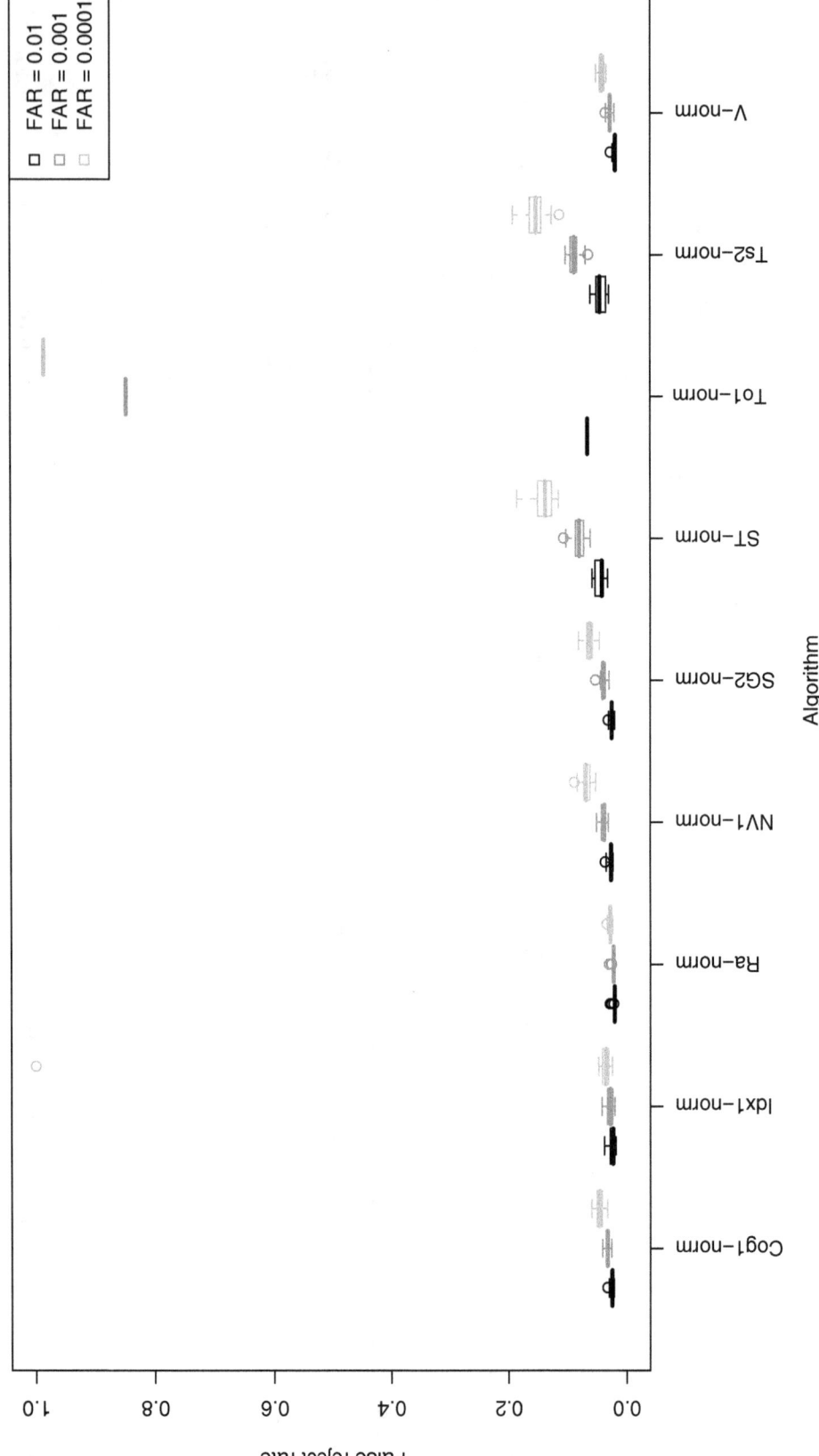

Figure 18: Results for controlled experiment on the high-resolution dataset for normalized algorithms.

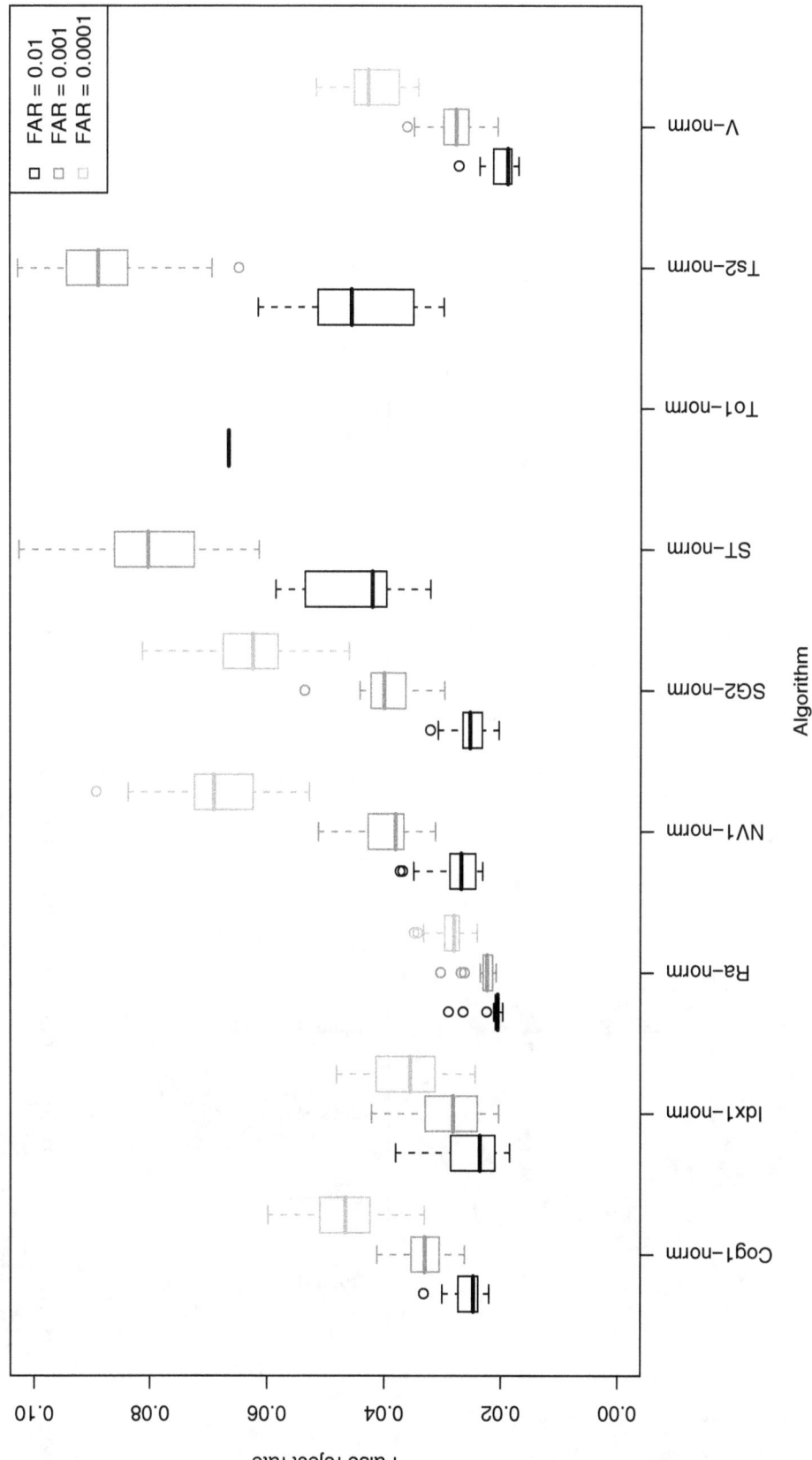

Figure 19: Results for controlled experiment on the high-resolution dataset for normalized algorithms. The range for FRR on the vertical axis is 0.00 to 0.10.

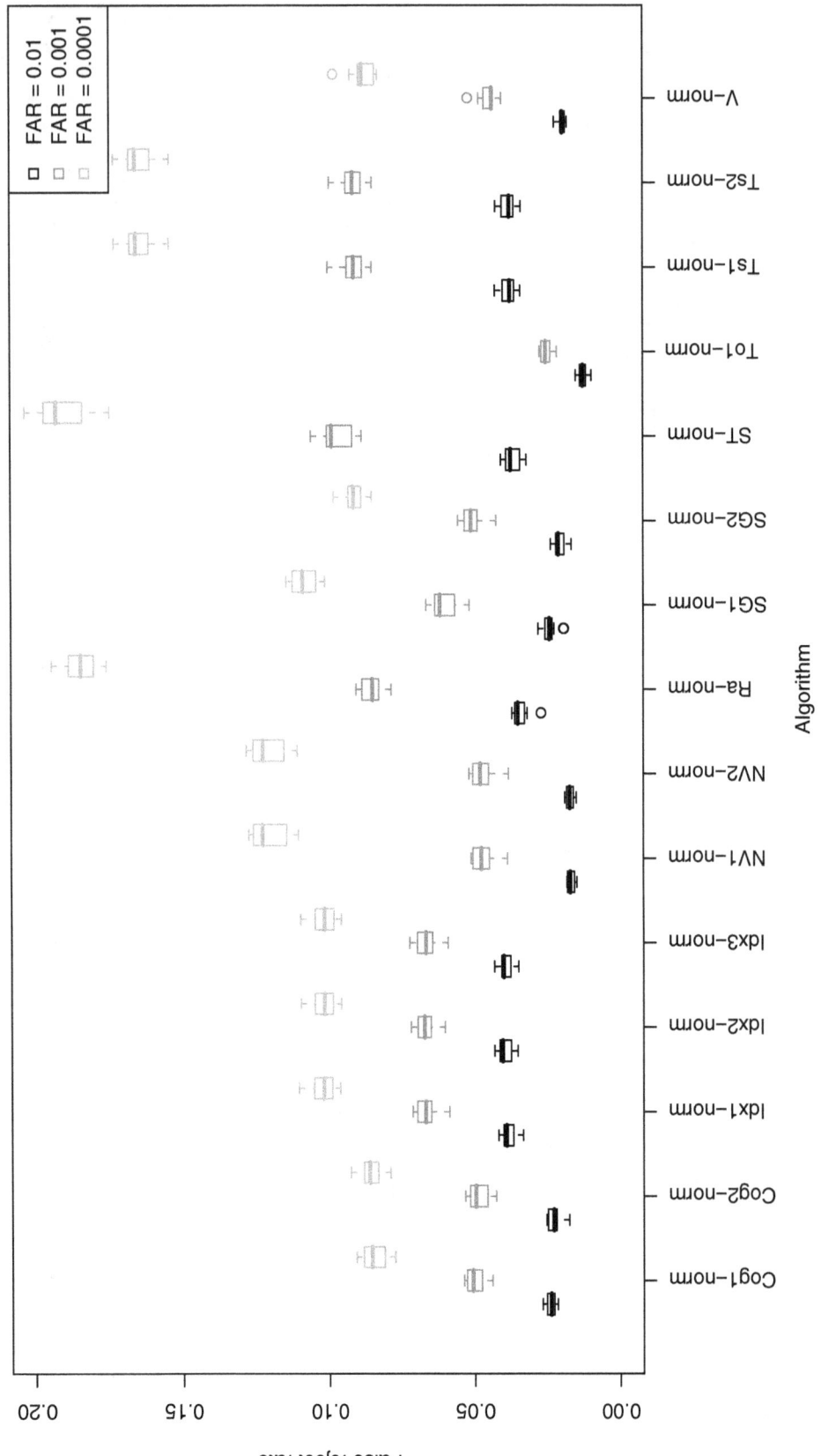

Figure 20: Results for controlled experiment on the the low-resolution dataset for normalized algorithms. The range for FRR on the vertical axis is 0.00 to 0.20.

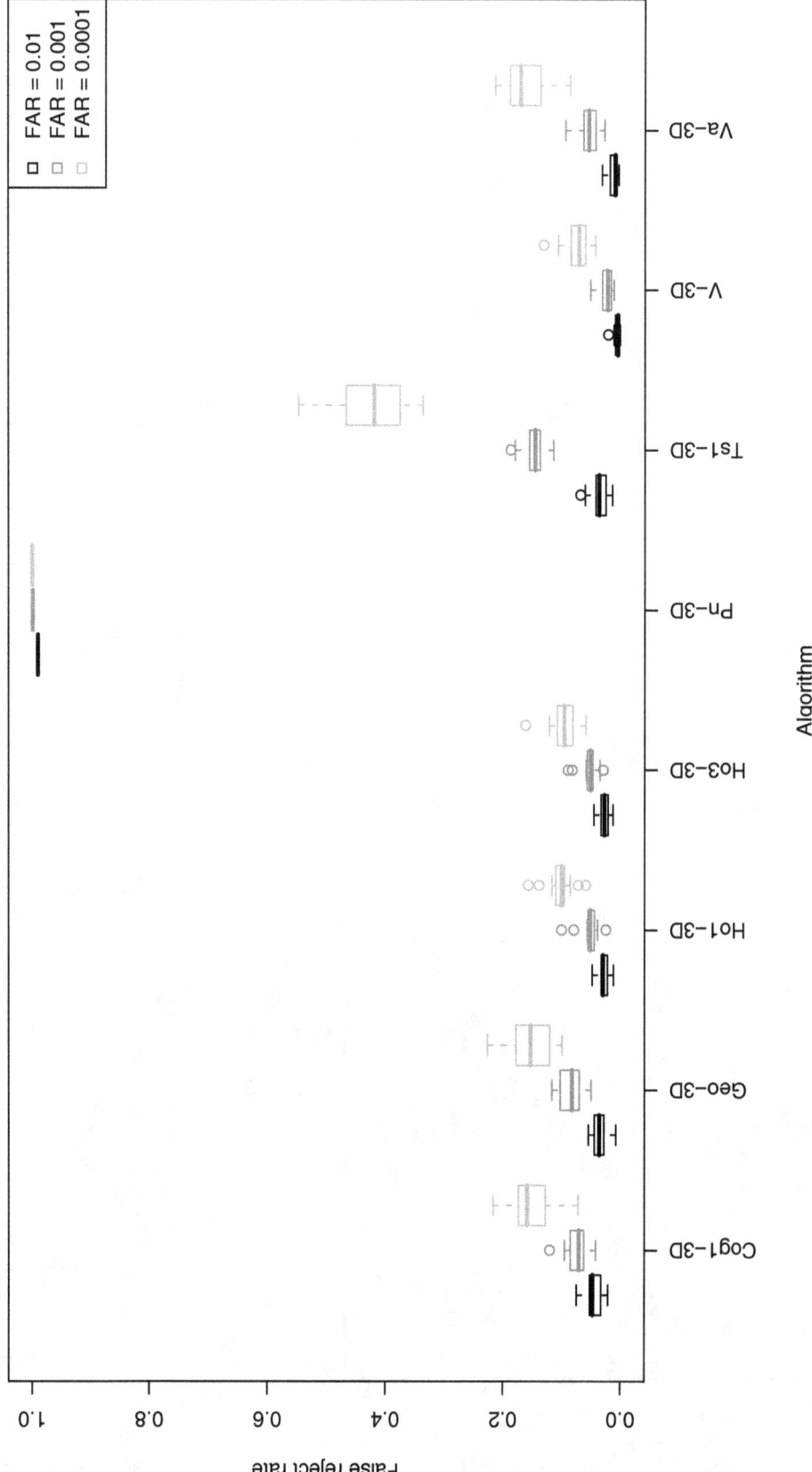

Figure 21: Results for 3D experiments for for one-to-one 3D algorithms

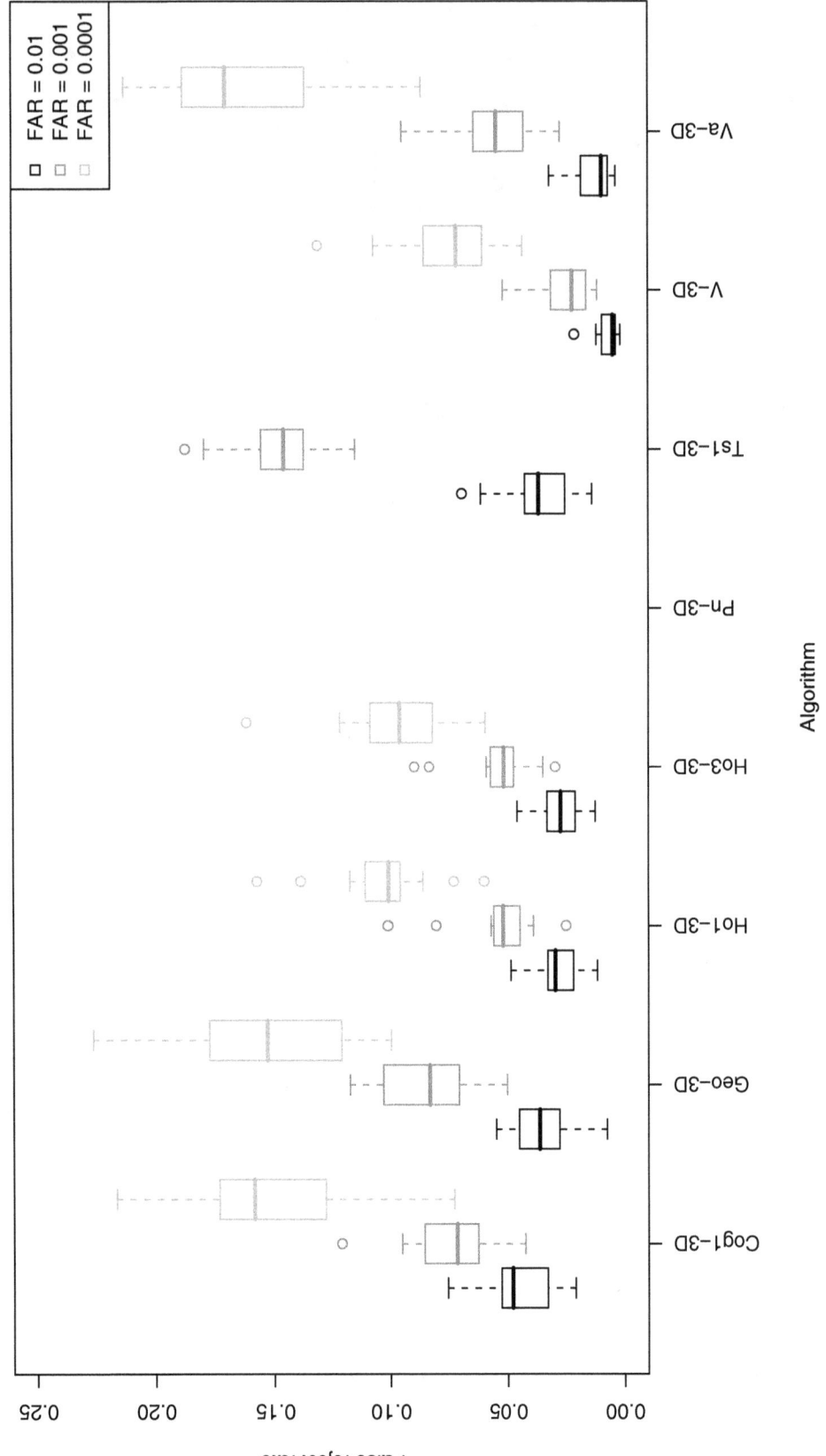

Figure 22: Results for 3D experiments for one-to-one 3D algorithms. The range for FRR on the vertical axis is 0.00 to 0.25.

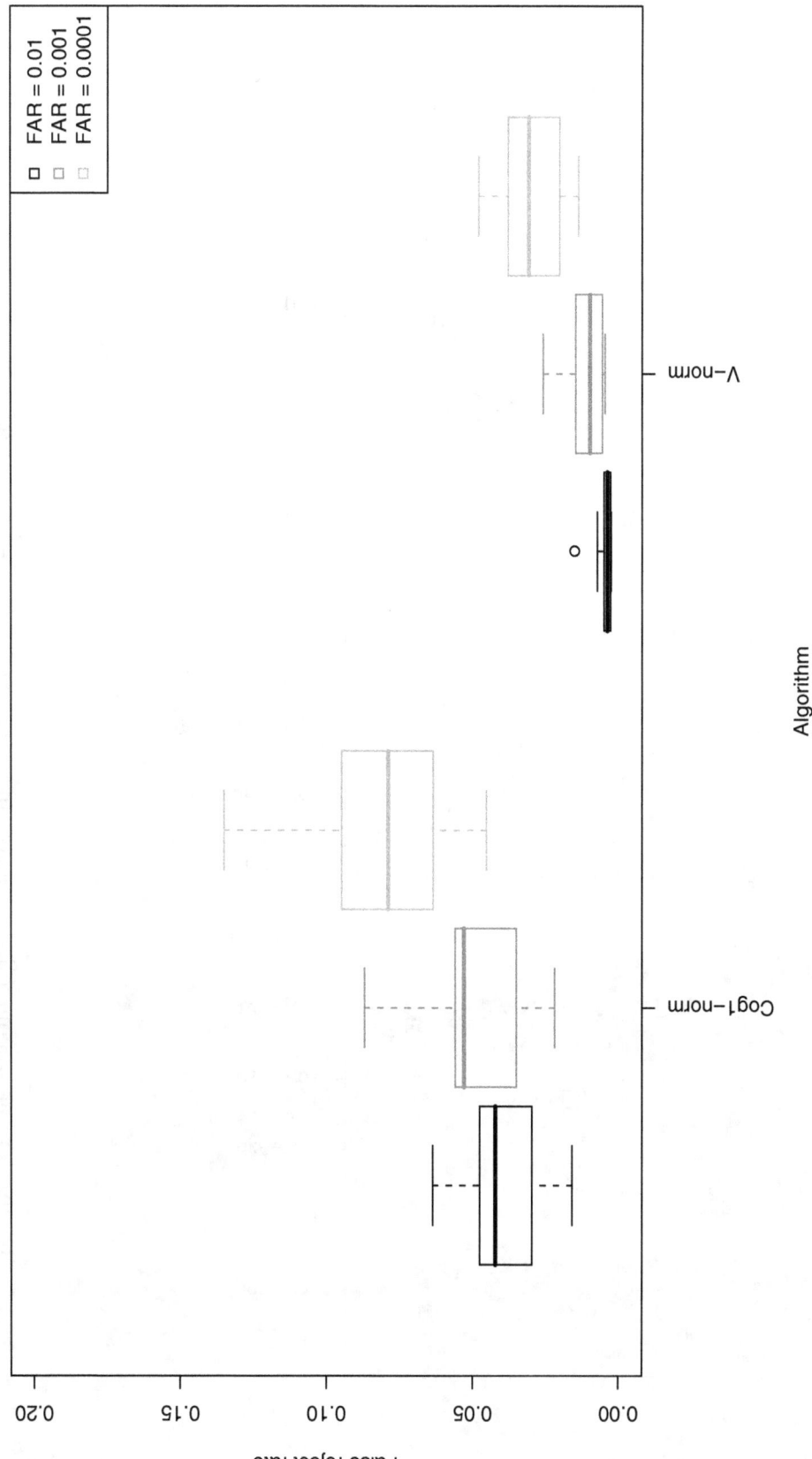

Figure 23: Results for 3D experiments for normalized 3D algorithms. The range for FRR on the vertical axis is 0.00 to 0.20.

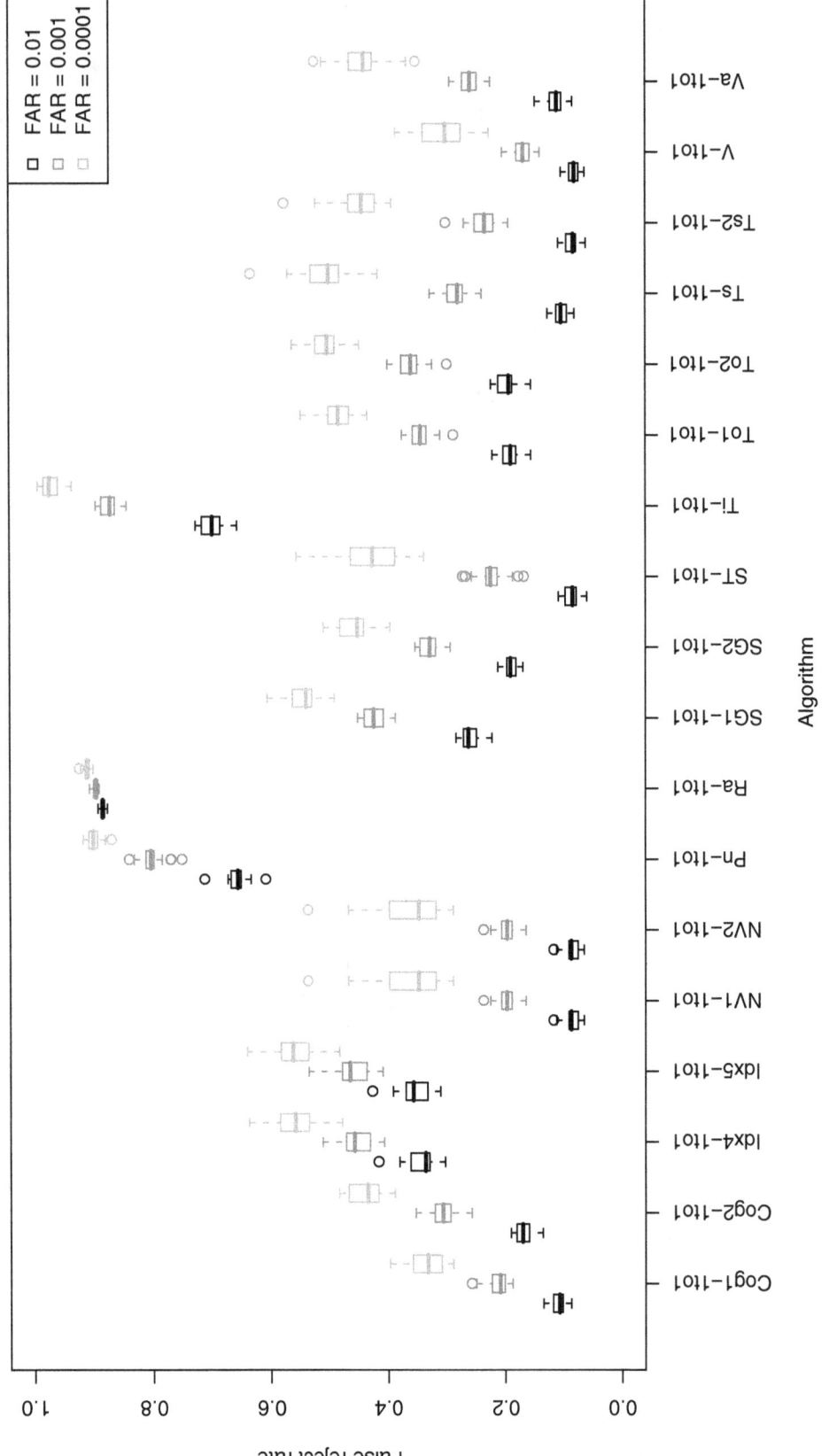

Figure 24: Results for the uncontrolled experiment on the very-high resolution dataset for one-to-one algorithms.

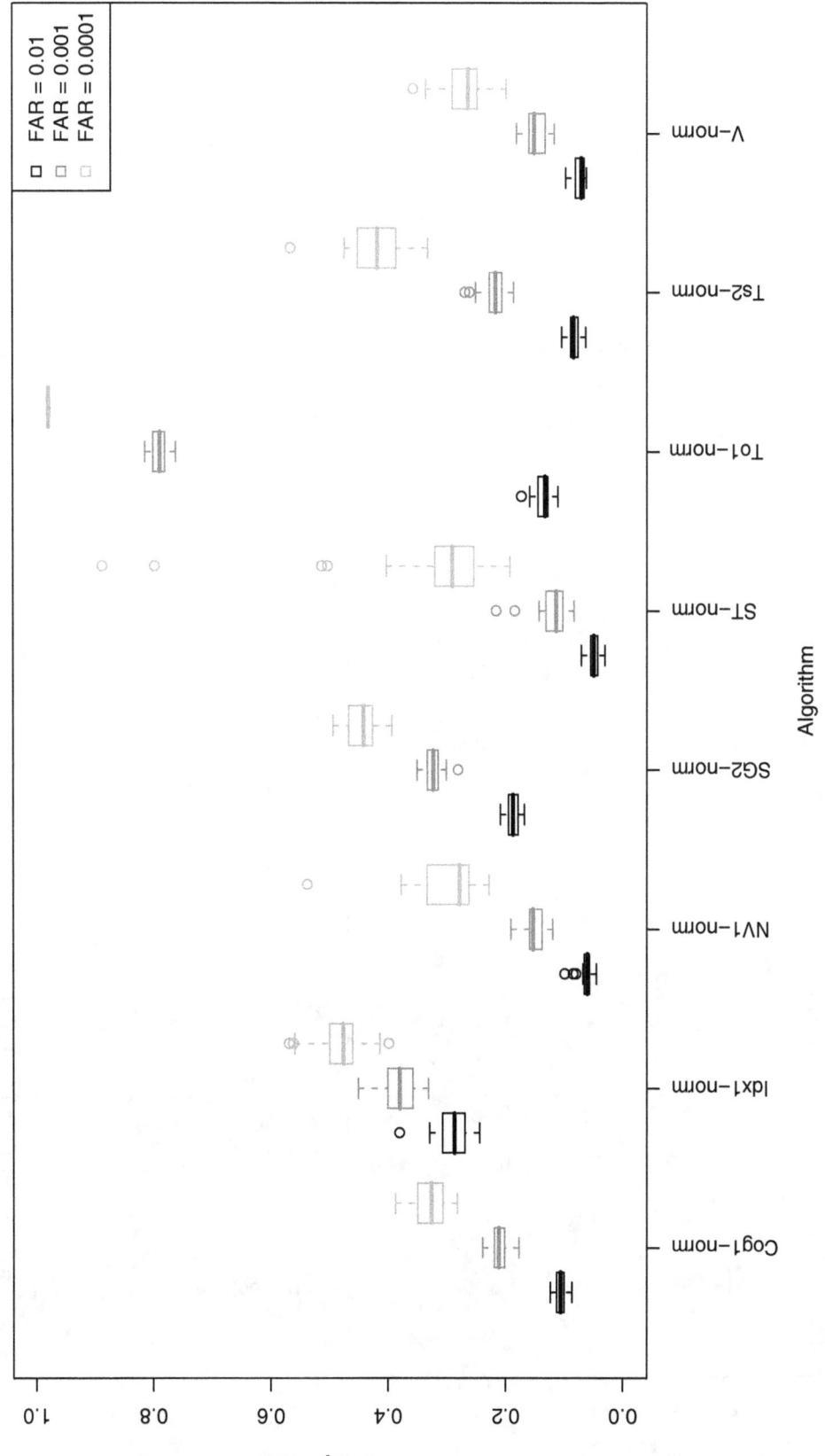

Figure 25: Results for the uncontrolled experiment on the very-high resolution dataset for normalized algorithms.

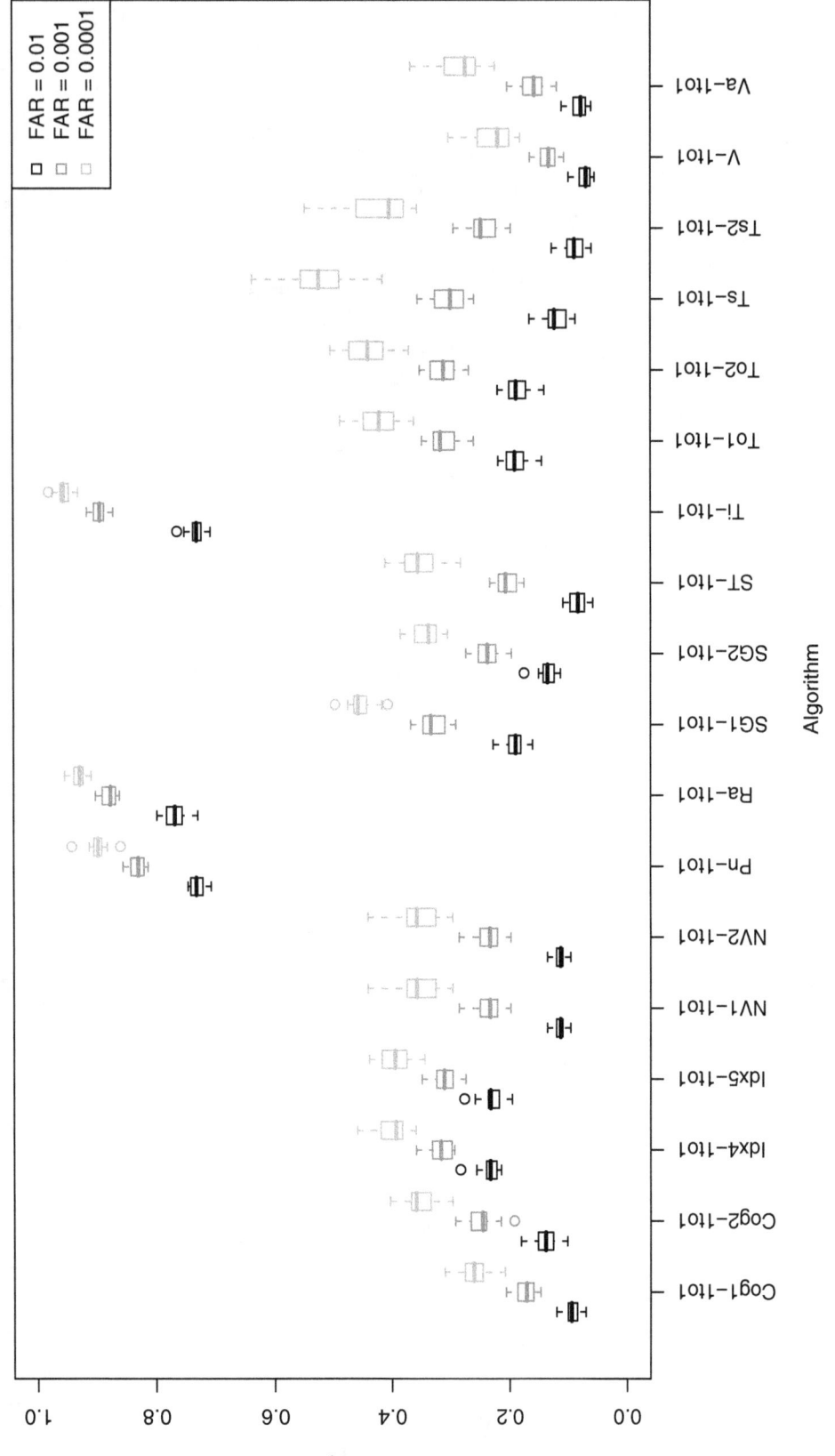

Figure 26: Results for the uncontrolled experiments on the high-resolution dataset for one-to-one algorithms.

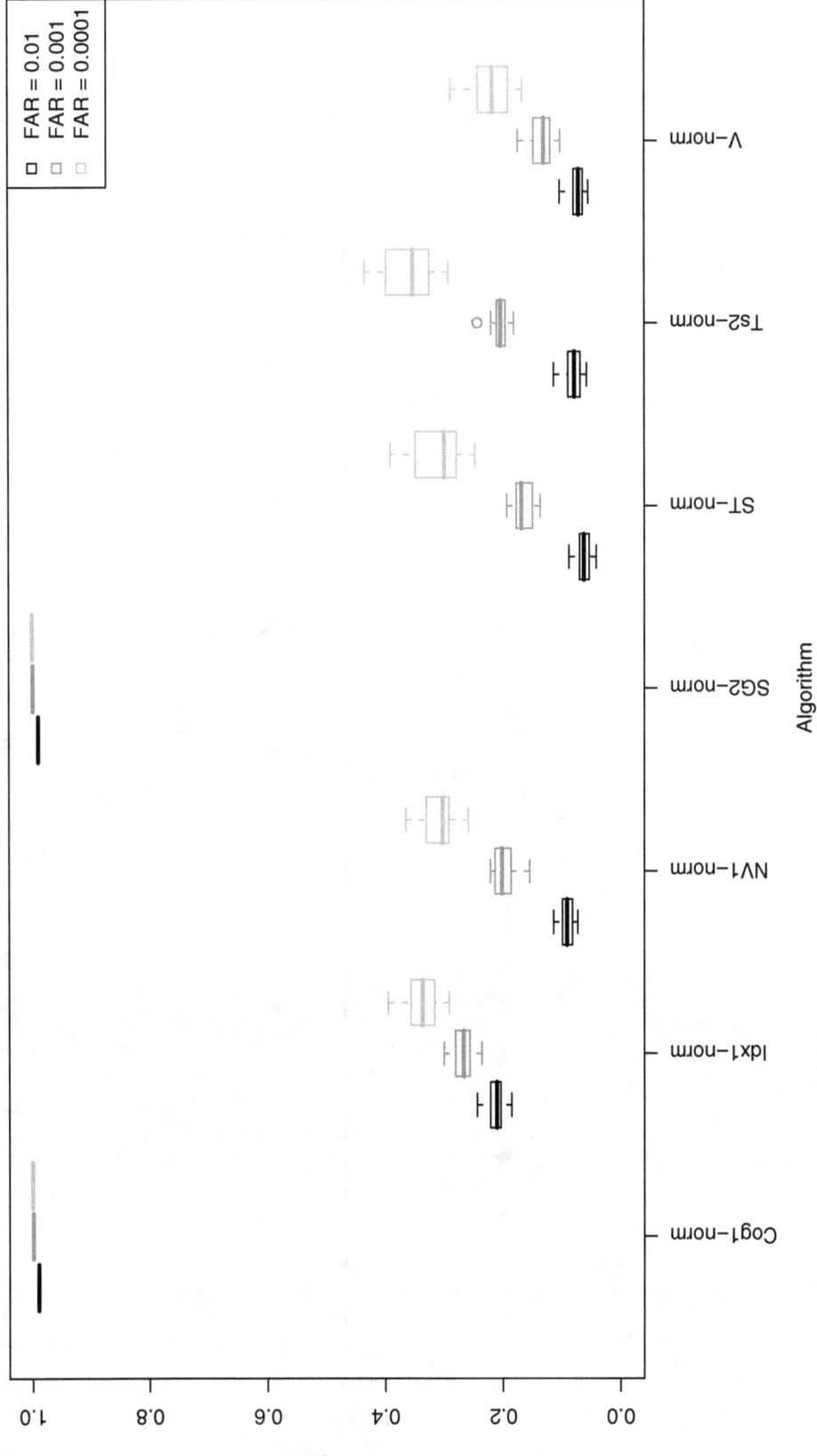

Figure 27: Results for the uncontrolled experiment on the high-resolution dataset for normalized algorithms.

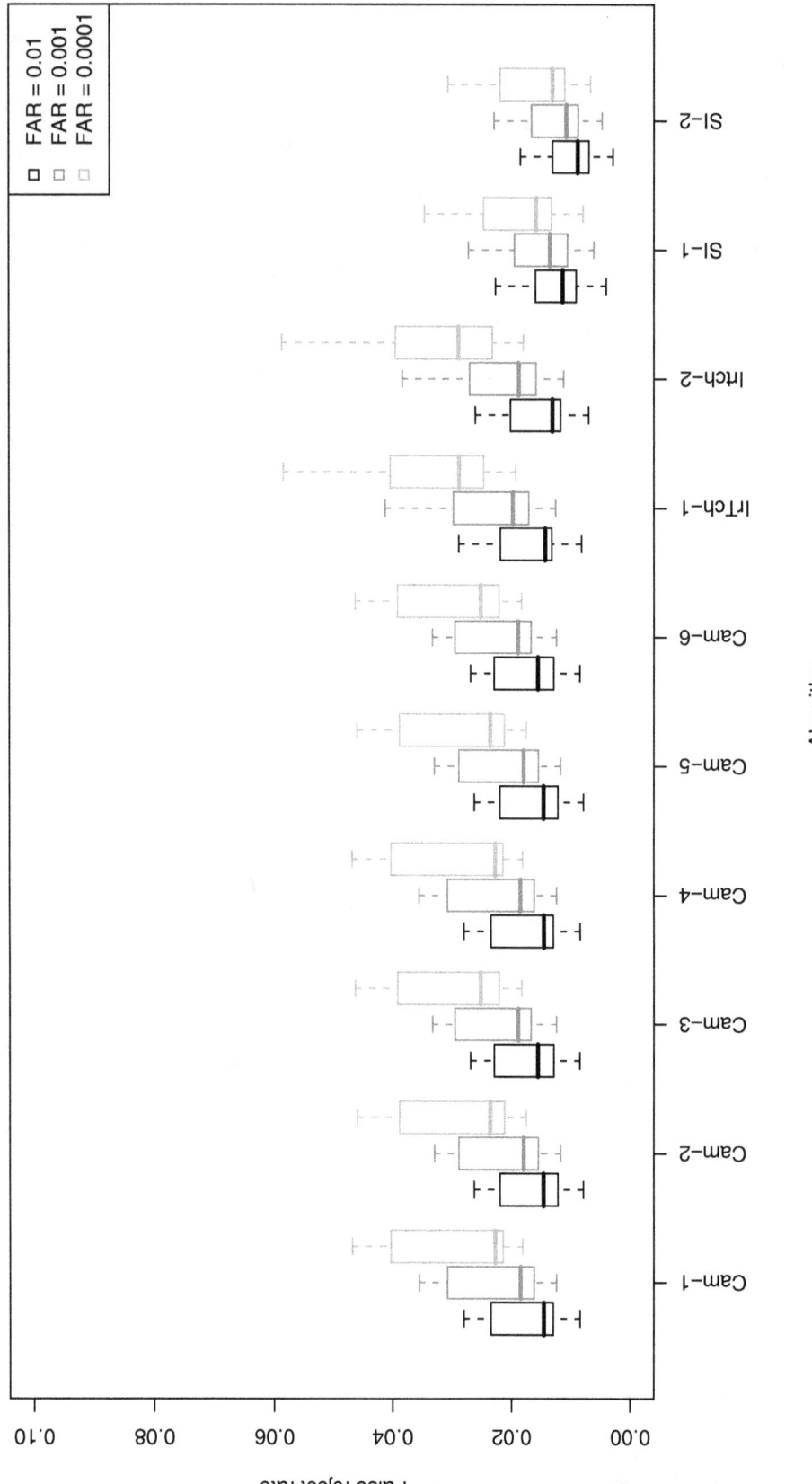

Figure 28: Results for the left iris experiment for single-iris algorithms. The range for FRR on the vertical axis is 0.00 to 0.10.

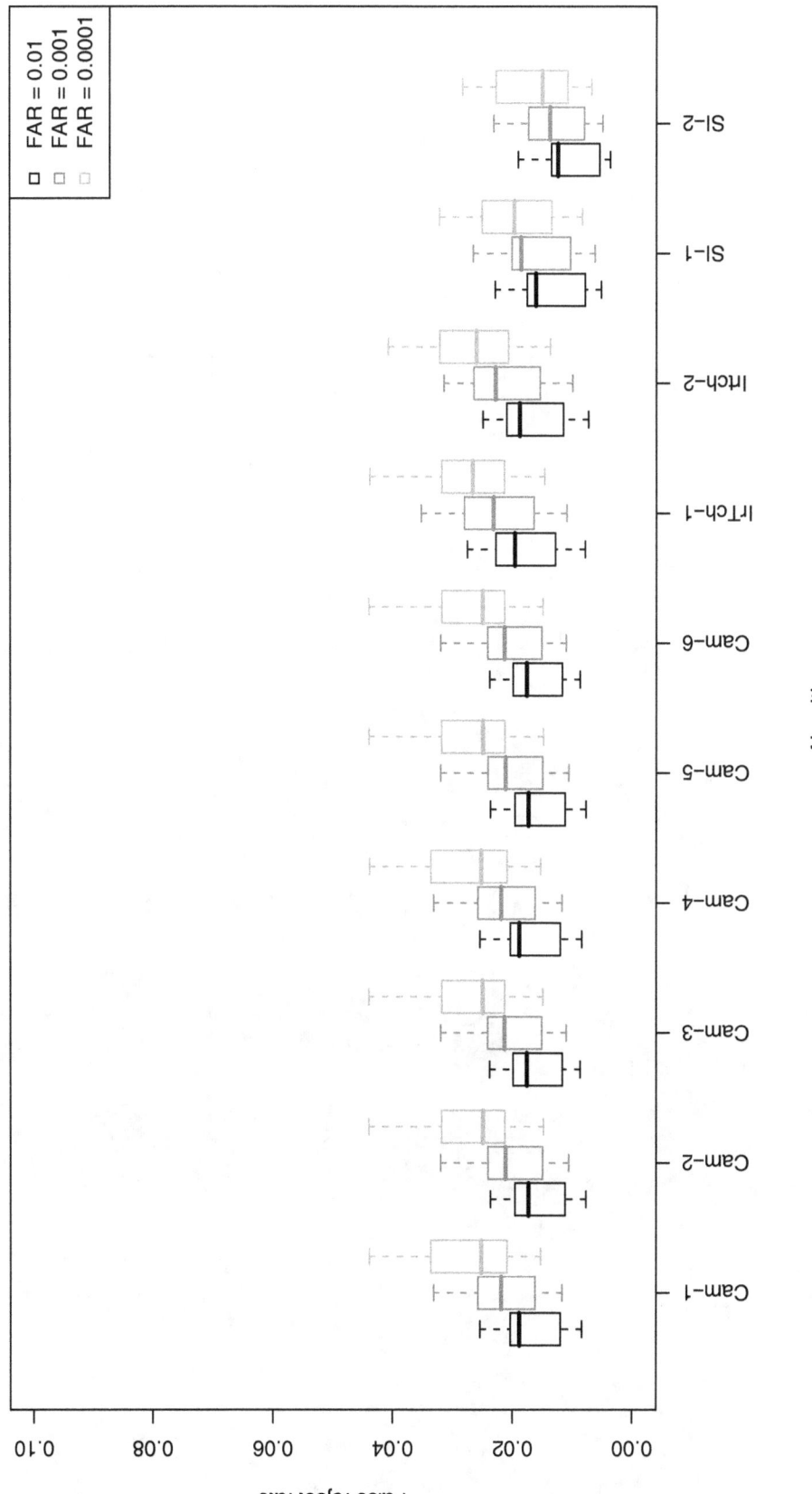

Figure 29: Results for the right iris experiment for single-iris algorithms. The range for FRR on the vertical axis is 0.00 to 0.10.

www.ingramcontent.com/pod-product-compliance
Lightning Source LLC
Chambersburg PA
CBHW081740170526
45167CB00009B/3891